# IMPLEMENTING THE CLIMATE REGIME

# IMPLEMENTING THE CLIMATE REGIME

## International Compliance

Edited by

Olav Schram Stokke, Jon Hovi and Geir Ulfstein

First published by Earthscan in the UK and USA in 2005

For a full list of publications please contact:
Earthscan
2 Park Square, Milton Park, Abingdon, Oxfordshire OX14 4RN
711 Third Avenue, New York, NY 10017

First issued in paperback 2016

*Earthscan is an imprint of the Taylor & Francis Group, an informa business*



Typesetting by The Fridtjof Nansen Institute
Cover design by Danny Gillespie

ISBN 13: 978-1-138-99240-5 (pbk)
ISBN 13: 978-1-84407-161-6 (hbk)

A catalogue record for this book is available from the British Library

Library of Congress Cataloging-in-Publication Data

Implementing the climate regime : international compliance / edited by Olav Schram Stokke, Jon Hovi, Geir Ulfstein.
p. cm.
Includes bibliographical references and index.
ISBN 1-84407-161-8 (hardback)
1. United Nations Framework Convention on Climate Change (1992). Protocols, etc., 1997 Dec. 11. 2. Climatic changes. 3. Climatic changes–Government policy. 4. Greenhouse gas mitigation–International cooperation. I. Stokke, Olav Schram, 1961- II. Hovi, Jon, 1956- III. Ulfstein, Geir, 1951-

QC981.8.C5I457 2005
363.738'74526–dc22

2004022850

# Contents

## Figures, Tables and Boxes

### *Figures*



### *Tables*



### *Boxes*

# Preface

Unlike most books on climate politics, this volume has a sharp focus on one particular aspect of the global climate regime – the system set up to improve compliance with commitments under the 1997 Kyoto Protocol. The design of a compliance system has been controversial ever since the adoption of the 1992 Framework Convention on Climate Change. The outcome, laid out in the 2001 Marrakesh Accords, is innovative in international environmental law and its operation will be decisive for the effectiveness of the climate regime.

This book has been prepared under a project coordinated jointly by three Norwegian research institutions targeting international climate politics: CICERO Center for International Climate and Environmental Research–Oslo, The Department of Public and International Law at the University of Oslo, and the Fridtjof Nansen Institute. The book also involves prominent scholars from the US. This transnational group of authors met for several workshops in the course of the project to ensure a high level of integration of the book's various chapters. Prior to the last workshop, the chapter drafts were subject to individual anonymous review by external experts in the various fields covered. We are very grateful to Scott Barrett, Thomas Gehring, Marc Levy, Sebastian Oberthür, Lasse Ringius, Detlef Sprinz and David Victor for generously offering advice on how the drafts could be further advanced.

In late 2003, the chapters were assembled into a coherent manuscript and submitted to James & James/Earthscan. We are highly appreciative of Earthscan's commissioning editor, Rob West, for his amiable efficiency in guiding the journey from manuscript to book. Thanks are due to the publisher's anonymous reviewers who provided insightful and constructive suggestions on individual chapters as well as the book structure. We would also like to thank M. J. Mace for her excellent editorial services, with respect to both contents and presentation. We are indebted to Maryanne Rygg for valuable editorial and practical assistance throughout the preparation of this book and to Rowan Davies and Lynn P Nygaard for editiorial services.

The project has been funded by the Samstemt Programme under the Research Council of Norway and by the participating institutions.

Oslo, June 2004

Olav Schram Stokke, Jon Hovi and Geir Ulfstein

# Notes on Contributors

**Steinar Andresen** is a professor at the Department of Political Science at the University of Oslo and a senior research fellow at the Fridtjof Nansen Institute, Norway. His research focuses on international management of natural resources and the environment with an emphasis on whaling, marine pollution, climate change and the role of the UN in global environmental governance. Analytically, Andresen has examined factors that influence the effectiveness of international cooperation, such as the interaction between science and policy, processes of leadership and specific institutional features. Andresen has published books, mostly with co-authors, for the Scandinavian University Press, Belhaven, Manchester University Press and MIT, and a number of articles in such journals as *Marine Policy*, *International Environmental Agreements*, *Global Environmental Politics* and *Global Governance*.

**Terje Berntsen** holds a Dr Scient degree from the University of Oslo and is a senior research fellow at CICERO Center for International Climate and Environmental Research–Oslo. He also holds a part-time position at the Department of Geosciences, University of Oslo. His research interest is modelling of atmospheric distribution, sources and sinks of greenhouse gases, air pollutants and aerosols. He has published a number of research articles in these fields in international peer-reviewed journals, including the *Journal of Geophysical Research*, *Tellus*, *Geophysical Research Letters* and *Climatic Change*. Berntsen has contributed to several of the previous assessment reports from Working Group 1 of the International Panel on Climate Change (IPCC), and has recently been elected as lead author for Chapter 2 of their fourth assessment report.

**Jan Fuglestvedt** holds a Dr Scient degree from the University of Oslo and is research director at CICERO Center for International Climate and Environmental Research–Oslo. His research interests include the role of atmospheric chemistry in the context of climate change, calculations of contributions to climate change, strategies for reducing man-made impacts on climate and application of global warming potentials (GWPs) and alternatives in climate policy. He has published a number of research articles in these fields in international peer-reviewed journals including *Tellus*, *Geophysical Research Letters*, *Ambio* and *Climatic Change*. Dr Fuglestvedt has also contributed to the assessment reports from Working Group 1 of the IPCC.

**Lars H. Gulbrandsen** is a research fellow at the Fridtjof Nansen Institute, Norway. His research interests are in the area of environmental politics, with a particular focus on international–domestic and public–private governance interactions. He has published articles in *Environmental Politics* and *Global Environmental Politics* and book chapters in edited volumes.

**Cathrine Hagem** is a post-doctorate student at the University of Oslo, Department of Economics. Her research interest is environmental economics, with special emphasis on the design of international climate agreements and domestic climate policy. Her research in this field includes theoretical studies of the impact of asymmetric information, market power and limited participation on the performance of climate agreements. She has also conducted theoretical and empirical work on the interplay between fossil fuel markets and markets for emission permits. Some of this work focuses on the links between strategic actions in the energy market and permit market. Her work has been published in such international journals as *The Energy Journal, Energy and Resource Economics, Environmental and Resource Economics* and the *Journal of Environmental Economics and Management.*

**Jon Hovi** is professor of political science, University of Oslo, and a senior researcher at CICERO Center for International Climate and Environmental Research–Oslo. His main research interest is the design and enforcement of international regimes. He is the author of *Games, Threats and Treaties: Understanding Commitments in International Relations* (Pinter, 1998).

**Ronald B. Mitchell** is an associate professor with tenure in the Department of Political Science, University of Oregon. He earned his PhD in public policy at Harvard University in 1992 and was a visiting associate professor at the Center for Environmental Science and Policy from June 1999 to December 2001. His publications include the award-winning book *Intentional Oil Pollution at Sea: Environmental Policy and Treaty Compliance* (MIT Press, 1994), articles in *International Organization, International Studies Quarterly* and *Global Governance,* and chapters in numerous edited volumes. His research focuses on the effectiveness of international institutions in influencing the behaviour of state and non-state actors, as well as on the influence of environmental science on international policy-making. He has been a member of the United States National Research Council's Committee on the Human Dimensions of Global Change and is currently a member of the editorial boards of *International Organization*, the *Journal of Environment and Development* and *Global Environmental Politics*. He teaches courses on international relations theory, international environmental politics and international regimes.

**Jon Birger Skjærseth** is a senior research fellow and research director at the Fridtjof Nansen Institute, Norway. His research interests are international environmental cooperation, national environmental policy and the strategies of non-state actors, particularly multinational companies. He has published extensively in these fields, including the books *North Sea Cooperation: Linking International and Domestic Pollution Control* (Manchester University Press, 2000), *Regime Effectiveness: Confronting Theory with Evidence* (MIT Press, 2001, with E. L. Miles, A. Underdal, S. Andresen, J. Wettestad and E. M. Carlin), *Climate Change and the Oil Industry: Common Problem, Varying Strategies* (Manchester University Press, 2003, with T. Skodvin) and *International Regimes and Norway's Environmental Policy: Crossfire and Coherence* (editor, Ashgate, 2004). He has also published extensively in international journals such as *Cooperation and Conflict*, *Environmental Politics*, *Global Environmental Politics*, the *Journal of Common Market Studies* and *Marine Policy*.

**Olav Schram Stokke** is a senior research fellow at the Fridtjof Nansen Institute, Norway. His research interest is international political economy, with special emphasis on regime theory, regional cooperation and international management of resources and the environment. He has published extensively in these fields, including the edited volumes *Governing High Sea Fisheries: The Interplay of Global and Regional Regimes* (Oxford University Press, 2001), *Governing the Antarctic: The Effectiveness and Legitimacy of the Antarctic Treaty System* (Cambridge University Press, 1996, with D. Vidas) and *The Barents Region: Cooperation in Arctic Europe* (SAGE, 1994, with O. Tunander). Stokke is co-editor of the *Yearbook of International Co-operation on Environment and Development* (Earthscan), and his work has been published in such international journals as *Marine Policy*, *Ocean Development and International Law*, *Ocean and Coastal Management* and the *Annals of the American Academy for Political and Social Science*. He is a member of the editorial board of *Global Environmental Politics*.

**Frode Stordal** is a professor in meteorology at the University of Oslo, Norway. He also holds an adjunct position at the Norwegian Institute for Air Research (NILU), Norway. He has a Dr Scient degree from the University of Oslo. His research interest is modelling and observing greenhouse gases, air pollutants and aerosols. He has published a number of research articles in these fields in international peer-reviewed journals, including the *Journal of Geophysical Research, Tellus, Geophysical Research Letters* and *Nature*. Stordal has contributed to several of the previous assessment reports from Working Group 1 of the IPCC. He has participated in more than 20 projects financed by the European Union (EU), and is currently coordinating two EU projects.

**Geir Ulfstein** is professor of law and director of the Norwegian Centre for Human Rights, University of Oslo. He has been deputy director at the Department of Public and International Law, University of Oslo. He has written extensively on the law of the sea, international environmental law and the law on the use of force. He is co-editor-in-chief of the *Yearbook of International Environmental Law* (Oxford University Press).

**Jacob Werksman** is senior adviser to the Global Inclusion Program of the Rockefeller Foundation, where he supports grant-making decisions in the area of international intellectual property rights policy and international trade policy. Prior to this, he served as environmental institutions and governance adviser to the United Nations Development Programme (UNDP). He joined UNDP after nearly ten years at the Foundation for International Environmental Law and Development (FIELD) where he served as a lawyer, programme director and, most recently, managing director. He is currently an adjunct professor of law at New York University, and was a lecturer in international economic law at the masters level at the University of London. He has held visiting academic posts at the United Nations University Institute of Advanced Studies, the University of Edinburgh, the University of Kent and at the University of Connecticut Law School. Before joining FIELD in 1992, Mr

Werksman practised energy and environmental regulatory law in California, where he remains an active member of the State Bar. He holds degrees from Columbia University (AB 1986, English Literature); the University of Michigan (Juris Doctor, *cum laude*, 1990); and the University of London (LLM, 1993, Public International Law).

**Hege Westskog** is a senior research fellow at CICERO Center for International Climate and Environmental Research–Oslo. An economist, her research interests are focused on the design of policy instruments. Specifically, she has worked on the design of an emission trading system to reduce emissions of climate gases. This includes both theoretical and empirical studies of strategic behaviour (the exercise of market power) in permit markets and the effects of different designs of intertemporal emission trading systems. She has also been involved in analysing the economic effects of different designs of climate agreements. Her work is published in journals such as the *Energy Journal*, the *Journal of Environmental Economics and Management* and *Environment and Development Economics*.

**Jørgen Wettestad** is a senior research fellow at the Fridtjof Nansen Institute, Norway. His research interest is regime theory and EU politics, with a special focus on questions related to the effectiveness and design of international environmental institutions. His particular empirical focus is on air pollution and climate change. He has published extensively, most recently: *Clearing the Air: European Advances in Tackling Acid Rain and Atmospheric Pollution* (Ashgate, 2002); *Environmental Regime Effectiveness: Confronting Theory with Evidence* (MIT Press, 2002, with E. L. Miles, A. Underdal, S. Andresen, J. B. Skjærseth and E. M. Carlin); *Science and International Environmental Regimes: Combing Integrity with Involvement* (Manchester University Press, 2000, with S. Andresen, T. Skodvin and A. Underdal); and *Designing Effective Environmental Regimes: The Key Conditions* (Edward Elgar, 1999). He is member of the Editorial Advisory Board of *Climate Policy*.

# Acronyms and Abbreviations

| | |
|---|---|
| AAUs | assigned amount units |
| ALTENER | an EU programme/decision on renewable energies |
| AOSIS | Alliance of Small Island States |
| API | American Petroleum Institute (US) |
| BAU emissions | business-as-usual emissions |
| BP | British Petroleum |
| BTU | British thermal unit |
| CAN | Climate Action Network |
| CDM | Clean Development Mechanism |
| CEE | Central and Eastern European (countries) |
| CEITs | countries with economies in transition |
| CEO | Chief Executive Officer |
| CERs | certified emission reductions |
| CFCs | chlorofluorocarbons |
| $CF_4$ | carbon tetrafluoride |
| $CH_4$ | methane |
| CIEL | Center for International Environmental Law |
| CITES | Convention on International Trade in Endangered Species of Wild Fauna and Flora |
| CLRTAP | Convention on Long-range Transboundary Air Pollution |
| CO | carbon monoxide |
| $CO_2$ | carbon dioxide |
| $CO_2e$ | carbon dioxide equivalent |
| COP | Conference of the Parties |
| COP/MOP | Conference of the Parties serving as the Meeting of the Parties |
| EB | Executive Body |
| CTM | chemical transport model |
| ECCP | European Climate Change Programme |
| ECE | Economic Commission for Europe |
| ECJ | European Court of Justice |
| ED | Environmental Defense (US |
| EEA | European Economic Area |
| EFTA | European Free Trade Association |
| EMEP | Cooperative Programme for Monitoring and Evaluation of the Long-Range Transmission of Air Pollutants in Europe |
| EMIT Group | Group on Environmental Measures and International Trade |

| | |
|---|---|
| ERTs | Expert Review Teams |
| ERUs | emission reduction units |
| ESA | EFTA Surveillance Authority |
| EU | European Union |
| EUROPIA | European Petroleum Industry Organization |
| FCCC | Framework Convention on Climate Change |
| FIELD | Foundation for International Environmental Law and Development (UK) |
| FoE | Friends of the Earth |
| GATS | General Agreement on Trade in Services |
| GATT | General Agreement on Tariffs and Trade (1947) |
| GCC | Global Climate Coalition |
| GEF | Global Environment Facility |
| GEMS | Global Environmental Monitoring System |
| GHG | greenhouse gas |
| GOOS | Global Ozone Observing System |
| GRULAC | Group of Latin American and the Caribbean (countries) |
| GWPs | global warming potentials |
| HFCs | hydrofluorocarbons |
| ICCAT | International Commission for the Conservation of Atlantic Tunas |
| ICJ | International Court of Justice |
| IGO | intergovernmental organization |
| IM | inverse modelling (or inversion modelling) |
| IPCC | Intergovernmental Panel on Climate Change |
| IWC | International Whaling Commission |
| JI | joint implementation |
| JUSCANZ | loose Annex 1 grouping including Japan, US, Switzerland, Canada, Australia, Norway and New Zealand |
| JWG | Joint Working Group (on Compliance) |
| LULUCF | land use, land-use change and forestry |
| MARPOL | International Convention for the Prevention of Pollution from ships |
| MCP | multilateral consultative process |
| MEA | multilateral environmental agreement |
| MMPA | Marine Mammal Protection Act (US) |
| $N_2O$ | nitrous oxide |
| NEPP | National Environmental Policy Plan (the Netherlands) |
| NGOs | non-governmental organizations |
| NILU | Norwegian Institute for Air Research |
| $NO_x$ | nitrogen oxides |
| NRDC | Natural Resources Defense Council (US) |
| $O_3$ | tropospheric ozone |
| ODSs | ozone-depleting substances |
| OECD | Organization for Economic Cooperation and Development |
| OH | hydroxyl |

| | |
|---|---|
| OPEC | Organization of Petroleum-Exporting Countries |
| PBL | planetary boundary layer (the lower 50–2000m of the atmosphere) |
| PFCs | perfluorocarbons |
| ppbv | parts per billion by volume |
| PPM | process and production method |
| ppmv | parts per million by volume |
| pptv | parts per trillion by volume |
| QELRCs | quantified emission limitation or reduction commitments |
| RF | radiative forcing |
| RMUs | removal units |
| SAVE | Specific Actions for Vigorous Energy Efficiency (an EU energy efficiency framework directive) |
| SBSTA | Subsidiary Body for Scientific and Technological Advice |
| SCM | Subsidies and Countervailing Measures (WTO) |
| $SF_6$ | sulphur hexafluoride |
| SIDS | small island developing states |
| $SO_2$ | sulphur dioxide |
| SOGE | System for Observation of halogenatial Greenhouse gases in Europe |
| SPS | Sanitary and Phytosanitary Measures (WTO) |
| STEP System | Shell Tradable Emissions Permit System |
| TBT | Technical Barriers to Trade (WTO) |
| TRIPS | Trade-related Aspects of Intellectual Property Rights |
| UKAS | United Kingdom Accreditation Service |
| UNCED | United Nations Conference on Environment and Development |
| UNCLOS | United Nations Convention on the Law of the Sea |
| UNDP | United Nations Development Programme |
| UNEP | United Nations Environment Programme |
| UNFCCC | UN Framework Convention on Climate Change |
| UNICE | Union of Industrial and Employer's Confederation of Europe |
| VOCs | volatile organic compounds |
| WBCSD | World Business Council for Sustainable Development |
| WEOG | Western European and Others Group |
| WMO | World Meteorological Organization |
| WRI | World Resources Institute (US) |
| WSSD | World Summit on Sustainable Development |
| WTO | World Trade Organization |
| WWF | World Wide Fund for Nature |

# Introduction and Main Findings

Jon Hovi, Olav Schram Stokke and Geir Ulfstein

## Purpose and plan

The climate regime now emerging is the product of a process that spans more than two decades. In 1979, the first World Climate Conference was held under the auspices of the World Meteorological Organization (WMO). Together with the United Nations Environment Programme, the WMO established the Intergovernmental Panel on Climate Change (IPCC) in 1988, to assess peer-reviewed scientific literature relevant for understanding the risk of human-induced climate change.

At the UN Conference on Environment and Development (UNCED) held in Rio in 1992, the UN Framework Convention on Climate Change (UNFCCC) was signed. Here, developed countries committed themselves to the adoption of policies and measures that would limit man-made greenhouse gas emissions and protect and enhance sinks and reservoirs within their territories, with the aim of stabilizing greenhouse gas emissions at their 1990 level by the year 2000.[1] These substantive commitments were 'soft' in that their wording was either aspirational, vague or legally non-binding. The form of these commitments is reflected in the UNFCCC's provision for a soft compliance mechanism in the form of a 'multilateral consultative process' (MCP)[2] – a body whose deliberations were to be conducted in a 'facilitative, cooperative, non-confrontational, transparent and timely manner, and be non-judicial'.[3]

The Kyoto Protocol, agreed in 1997, represented a large step forward. The protocol prescribes quantified limits and reduction obligations on each Annex I party's (developed country's) average greenhouse gas emissions over a first commitment period, to run from 2008 to 2012, aimed at reducing the overall emissions of such gases by at least 5 per cent below 1990 levels. A number of 'flexible mechanisms' were created to enable the parties to achieve these reductions in a cost-effective way.[4]

These binding and ambitious commitments by developed states were later supplemented by a compliance system, elaborated in the 2001 Marrakesh Accords, which establish a Compliance Committee with both a Facilitative Branch and an Enforcement Branch. While the Facilitative Branch will provide 'advice and facilitation to Parties in implementing the Protocol', the Enforcement Branch will determine whether a party is in non-compliance with its emission target and its reporting requirements, and determine whether it satisfies eligibility requirements for participation in the flexible mechanisms.[5] The Marrakesh Accords also specify a list of 'consequences' to be imposed by the Enforcement Branch in the second commitment period (2013–2017) on countries which fail to comply with their first period commitments.

This book explores the nature and effectiveness of the climate regime's compliance system. The purpose of this introduction is to pinpoint the main findings of the chapters that follow. We begin by describing in more detail the main features of the compliance system and how it came about. We then go on to ask, in light of the subsequent chapters, how effective the Kyoto–Marrakesh compliance system is likely to be. Can we expect this system to help parties overcome challenges related to effective verification, review and response? We also ask if external means of enforcement have a role to play. Are there good reasons to restrict the use of external enforcement? What relevance can trade sanctions have in the climate context? Against this background, we end by discussing the criticisms that have been made of Kyoto's compliance system, and what these criticisms imply for the system's ability to function in practice.

## Kyoto's compliance system: emergence and design

Any international compliance system must address at least three tasks: the verification of factual information about compliance, the review of this factual information in light of the legal commitments that states have assumed, and the identification of an appropriate response to violations of commitments.[6]

*Verification* entails an assessment of the completeness and accuracy of compliance-related information and its conformity with pre-established standards for reporting (Loreti et al, 2001, p3). As is the case with most multilateral environmental agreements (MEAs), verification under the Kyoto Protocol is primarily based on reporting by the parties themselves. A special feature of the climate regime is, however, the establishment of Expert Review Teams (ERTs), whose task it is to conduct a 'thorough and comprehensive technical assessment' of the performance of each Annex I party to the protocol; see Chapter 2 by Ulfstein and Werksman.

The ERTs are to provide independent information to the *review* process and identify 'questions of implementation'. However, they do not have the competence to determine non-compliance in a party's performance. While in other MEAs the Conference of the Parties determines cases of non-compliance, the Marrakesh Accords delegate these powers to the Enforcement Branch of the Compliance Committee.

With respect to the third task of a compliance system, response to violations of commitments, we may usefully distinguish between *facilitation*, aimed at enhancing the capacity of states to comply with their commitments, and *enforcement*, aimed at deterring future non-compliance. Under the Marrakesh Accords, the Facilitative Branch is to provide advice and facilitate the transfer of technological or financial resources to parties shown to be in non-compliance and to countries with a potential for non-compliance. The Enforcement Branch is to determine the response to non-compliance in the form of enforcement 'consequences'. For a failure to meet quantified greenhouse gas limitation or reduction commitments, these consequences are: (i) a deduction from the party's assigned amount for the second commitment period of 1.3 times the amount of that party's excess emissions in the first commitment period; (ii) a requirement that the non-compliant party develop a compliance action plan; and (iii) suspension of the party's eligibility to engage in transfers under the protocol's emissions trading provisions until that right is reinstated.[7]

In prescribing the use of such radical enforcement measures, the parties have also been determined to include safeguards against the abuse of such powers. First, the parties have agreed to promote automaticity in enforcement, rather than entrust the Enforcement Branch with discretion with regard to possible enforcement consequences. Accordingly, the Marrakesh Accords clearly establish which enforcement consequences shall attach to which violations.

Second, the compliance system's extensive powers to impose enforcement consequences have necessitated a strong focus on the establishment of a fair and credible institutional regime. The Compliance Committee and its Enforcement Branch are composed of a nominated group of independent experts, who reflect an agreed balance between Annex I and non-Annex I countries, while also taking into account representation from different regions of the world.

Third, the procedural protection of the party whose compliance is under consideration is ensured in several ways. The party in question has the right to be represented before the Branch, to have access to any information provided by others, and to comment on such information. The Compliance Committee's decision to allocate a question of implementation to the Enforcement Branch, rather than the Facilitative Branch, will be taken by a majority of the four Bureau members, two of whom are from Annex I Parties and two from non-Annex I parties. For the Enforcement Branch to decide on whether to proceed with a case after a preliminary examination, and to make a determination of non-compliance, a three-quarters majority is required and must include a majority of both Annex I party members and non-Annex I party members.[8] A party found to be in non-compliance may appeal a non-compliance decision to the Meeting of the Parties, if that party believes it has been denied due process. The decision of the Enforcement Branch may then only be overridden by a three-quarters majority vote of the MOP, which then refers the matter back to the Enforcement Branch.[9]

While these institutional and procedural features run parallel to the due process elements of domestic courts, the bipartisan character of civil courts and the prosecution aspect of a criminal case are avoided. We discuss below to what extent this system can be expected to make effective decisions.

The parties have not only been concerned with the content of the enforcement consequences and the institutional and procedural framework for their adoption, but also the legal status of those consequences. While legally binding consequences might be assumed to be more effective in ensuring compliance, Article 18 of the Kyoto Protocol provides that procedures and mechanisms entailing binding consequences can only be adopted by the protocol's cumbersome amendment procedure. Ulfstein and Werksman argue in Chapter 2 that whether or not a consequence should be considered binding under Article 18 may depend on which consequences are examined. Some enforcement measures, such as the suspension of eligibility to participate in the flexible mechanism, could be considered part of the 'implied powers' of international institutions. Others, such as the deduction of tonnes of a party's assigned amount at a penalty rate in a subsequent commitment period, are arguably of such a nature as to require an amendment of the protocol. The lack of clarity on these issues may result in an unclear political situation after the first commitment period, for example if certain Annex I parties that are found to be in non-compliance have not ratified an amendment that reflects the binding nature of these penalties. Indeed, those countries which find it most difficult to fulfil their obligations are also the least likely to ratify such an amendment (Barrett, 2003).

As Werksman describes in Chapter 1, the developed countries took a leading role in designing the Marrakesh compliance system, relying in the negotiations on their domestic experience and legal culture. Although many developing countries had opposed the establishment of a strong compliance system in the negotiation of the Framework Convention (Werksman, 1996, p95), they were united in their support of a strong enforcement mechanism under the Kyoto Protocol – but they insisted that it should only apply to Annex I parties. Somewhat paradoxically, while the US succeeded in achieving agreement on its favoured elements, including predictable and clearly defined enforcement consequences, it chose not to ratify the Protocol. The more reluctant Annex I states – Japan, Australia and the Russian Federation – succeeded in preventing a clear definition of the legal status of the system's enforcement measures. This proved to be one of the most contentious issues in the final rounds, and the question was ultimately left for the Meeting of the Parties to resolve.

## Challenges and prospects: an effective compliance system?

Like other key elements of the climate regime, such as the flexibility mechanisms, the climate compliance system is still in a formative phase. The Marrakesh Accords flesh out the broad contours provided by the Kyoto Protocol, but the institutions defined here have yet to constitute themselves and begin fulfilling their functions. Accordingly, an objection that it is too early to evaluate the system would not be unreasonable. On the other hand, while the proof of the value of the Kyoto compliance system lies in what it actually delivers, we might make a qualified judgement at this stage nonetheless, based on the system's characteristics. The fact that the Marrakesh Accords provide a rather detailed framework for the climate compliance system is exactly why the contributors to this book have volunteered their views on the likely effectiveness of this system.

The Kyoto compliance system is part of a broader international institution, the climate regime – and as students of regime effectiveness have shown, institutions can influence behaviour that is relevant to addressing international challenges, such as global warming, in several distinctive ways.[10] One way is to shape the incentives of parties by rendering non-compliance more costly or adherence with international norms more profitable. This 'logic of consequentiality' is emphasized by contributors to the so-called enforcement school in the study of international compliance (eg Barrett, 2003; Downs et al, 1996).[11] As argued by the so-called managerial school, however, the level of compliance with international agreements is generally quite good despite the fact that most international regimes have paid relatively little attention to enforcement (Chayes and Chayes, 1995). This conclusion has been supported by empirical studies of international environmental agreements (Brown Weiss and Jacobson, 1999; Jacobson and Brown Weiss, 1998). According to the Chayeses, the explanation is that compliance is generally *not* determined through deliberate decisions by states, made on the basis of expected costs and benefits. Instead, it is largely the result of norm-based behaviour and bureaucratic routines. Similarly, Franck (1990) emphasizes that compliance is the result – at least in part – of the relevant rule being perceived as legitimate by those to whom it is addressed.

These various processes and dynamics that underly the effectiveness of international regimes are relevant when examining the role of the Kyoto compliance system in overcoming challenges to effective verification, review and response in the climate context.

## *Verification: reliability of national reports*

A high quality system for gathering and verifying information on sources and sinks of greenhouse gases is a prerequisite for authoritative review and legitimate response. National reports on inventories of climate emissions are made according to detailed guidelines and reporting instructions developed by the IPCC and agreed upon collectively by regime members (see Chapter 4 by Berntsen and associates).[12] Compared to those of other international environmental regimes, the climate verification system has a fairly centralized component in the Expert Review Teams, whose members are chosen for their technical expertise and operate in their personal capacity. The main purpose of these teams is to ensure a high level of methodological consistency across national reports, including by means of adjusting national emissions inventories if these inventories are found to be inconsistent with the good practice guidance.

The impact of such centralization is to limit the leeway for opportunistic estimation of national emissions and removals. Given the complexity and variety of the activities that affect atmospheric concentrations of greenhouse gases, the climate verification system faces substantial challenges. Emissions from many sources will be measured by indirect methods – for instance, methane emissions on the basis of counts of types of livestock and acreage under cultivation, and vehicle carbon dioxide emissions on the basis of fuel consumption data (see Chapter 3 by Mitchell and Chapter 4 by Berntsen and associates). Uncertainties will abound in such estimations, and the appropriate conversion parameters will be debatable. For example, acreage emissions from managed ecosystems, such as nitrous oxide from agriculture, can be highly variable across geographic areas and over time.

On the other hand, the Kyoto rules pertaining to reporting on emissions and on other climate-related activities create not only obligations, but also an apparatus for enhancing the capacity of parties that have difficulties in meeting their reporting requirements. The Facilitative Branch of the Compliance Committee will serve as an institutional hub for support activities, which may involve technology transfer and verification capacity enhancement.[13] Such activities will presumably be particularly relevant for developing countries engaged in projects under the Clean Development Mechanism (CDM), but may also prove useful to Annex I economies in transition.

There is clearly a potential for drawing on experiences in other international regimes in the operation of the climate verification system, not least because many practitioners within the climate regime have been involved in the ozone and transboundary air pollution regimes (see Chapter 10 by Wettestad). In particular, experiences of the implementation committees under those regimes will prove valuable to the efforts of Expert Review Teams in improving the compilation and reporting of national climate data.

## *Alternative sources of information*

Despite its detailed guidelines, structures for capacity enhancement, and the possibility of adaptive learning, the climate verification system will be more effective if it can draw upon other sources of reliable data in addition to national reports. The Expert Review Teams are not explicitly obliged to seek and consider inputs from civil society organizations, but their mandate to provide a 'thorough and comprehensive' assessment suggests that they will do so. There are also clear indications that the opportunities in the Marrakesh Accords for non-governmental organizations (NGOs) and outside experts to provide relevant factual and technical information to the relevant branch of the Compliance Committee will actually be used.[14] Andresen and Gulbrandsen show in Chapter 8 that several large environmental organizations are determined to monitor the climate behaviour of key states and disseminate pertinent information by the means and channels available to them. Not only environmental organizations are likely to make use of these procedural openings; industrial companies, including those associated with production and distribution of fossil fuels, also have the incentive and the resources to feed the system with relevant information (see Chapter 9 by Skjærseth).

A more radical way to obtain independent checks on national reports – inverse modelling – has been used with considerable success in the management of transboundary air pollution. Inverse modelling is based on numerical transport models, observations of atmospheric concentrations and data on emissions from anthropogenic and natural sources (see Chapter 4 by Berntsen and associates). However, the multitude of sources and sinks for greenhouse gases, and the lack of representative measurements, limits the potential of this approach in the climate context, at least with the current density of the observational network. While at present inverse modelling cannot demonstrate with sufficient certainty that emissions from a given country are higher than those reported in its inventory, Berntsen and associates argue in Chapter 4 that these techniques may supplement the Marrakesh verification system by identifying cases that warrant closer examination.

## *Review: mixed blessings of the flexibility mechanisms*

The parties have agreed that they may make use of the so-called 'flexibility mechanisms' in meeting their commitments under the Kyoto Protocol. Through these mechanisms, parties may add to their assigned amounts, as defined by the limitation and reduction commitments under Annex B to the Protocol, assigned amount units aquired through international emissions trading, emission reduction units obtained through joint implementation, certified emission reduction units obtained through CDM projects, and removal units achieved through sink enhancement.[15]

The flexibility mechanisms are important to the cost effectiveness of the climate regime – the regime's ability to trigger mitigation or adaptation measures that minimize the wasteful use of scarce resources. Given the scale of the efforts required for reducing atmospheric concentrations of greenhouse gases, cost effectiveness is also essential to the social acceptability of the regime. However, the overall impact of the flexibility mechanisms on the effectiveness of the climate regime remains unclear, because with this increased flexibility comes the enhanced transactions costs associated with verification and review. We have noted already that large components of greenhouse gas emissions will be measured indirectly by means of conversion parameters that are vulnerable to challenge in the context of a non-compliance proceeding. Moreover, the emissions reductions or removals associated with joint implementation, the CDM or sink enhancement projects face considerable problems of causal substantiation, including resolving counterfactual questions of what level of emissions and removal would have occurred in the absence of those projects. This situation has led some observers to conclude that emission inventory systems will lack data of sufficient quality to permit determinations of non-compliance and subsequent enforcement measures. If this is the case, the 'carrots' of the climate regime could undermine the 'sticks'.

The verification and review challenges introduced by the flexibility mechanisms may reduce the legitimacy of climate commitments. Certainly, one fundamental contribution of a well functioning compliance system to a regime's legitimacy is to ensure that the violation of the fundamental norms of the regime triggers significant costs for the relevant party. Without such provisions, the creators of a regime signal that they do not consider these norms sufficiently important to justify constraints on how they are interpreted. To the extent that the verification and review challenges of the flexibility mechanisms undermine the linkage between violation and response, the ability of the climate rules to compel normative behaviour will suffer.

However, this may be an excessively pessimistic view. Mitchell argues in Chapter 3 that, over time, the bodies that operate the 'transaction log' system, track exchanges between countries and across periods, and certify and verify emissions reductions, may well develop baselines and monitoring procedures that are sufficiently accurate for the parties involved to accept the results. Many NGOs were adamantly opposed to the flexibility mechanisms during the negotiation of the Kyoto Protocol. Andresen and Gulbrandsen show in Chapter 8 that some green advocacy groups are presently developing the capacity to monitor and evaluate certified CDM and sink projects in order to close what they perceive as 'loopholes' of the climate regime. These efforts may serve to support the 'sticks' of the compliance regime, despite the added challenges of the flexibility mechanisms.

## *Naming and shaming*

Review bodies in environmental regimes are becoming more outspoken about non-compliance, and the naming and shaming of states whose performances fail to meet agreed standards is no longer unusual. It is worth noting that even before any formal review or response activities are undertaken, the information compiled under the climate verification system may trigger political costs to governments. The level of transparency of verification and compliance procedures is therefore important as it helps relevant information enter domestic political debates.

At the same time, climate transparency is not always negative for a government. By standardizing procedures and reducing the margin for opportunistic bookkeeping, the climate verification and compliance system permits governments to illustrate to their constituencies that they take the climate challenge, and compliance with related commitments, seriously.

Similarly, the enhanced availability of climate information makes it easier for civil society organizations to establish shaming infrastructures parallel to those of the climate regime itself. This may already be seen in websites and reports sponsored by large NGOs that warn against purchasing commodities from firms they deem as particularly climate-unfriendly. It is also seen in initiatives by NGOs to identify and shame buyers of 'hot air', i.e., traded assigned amount units that do not result from one state's mitigation efforts but rather from a decrease in that state's industrial production (and thus emissions) since the base year of the Kyoto targets.[16] The desire to use information to apply political pressure to governments and the private sector additionally explains the concern of environmental as well as industrial organizations when the first set of meetings under the CDM Executive Board provided for only very limited participation by non-governmental observers (Jacob, 2002).

At the same time, formal compliance with international commitments is not always sufficient grounds for a government to escape political criticism in a contested policy field; lawyers and the general public often use different metrics in evaluating non-compliance and responding to it. If a state's annual national greenhouse gas inventory displays a stepwise increase in national greenhouse gas emissions, domestic actors favouring more forceful climate measures are likely to present this as evidence that government policies lack the teeth required.

Finally, as described by Skjærseth in Chapter 9, greater attention to corporate climate accountancy provides opportunities for energy intensive companies to differentiate themselves from competitors by waving flags of climate friendship. This 'bragging' strategy is apparently one pursued by certain European petroleum companies, including British Petroleum (BP), Amoco and Shell. Such processes, and concomitant activities such as research on energy efficient production and consumption, will be important for the ability of states to meet their climate commitments. That said, there are clear limits to the force of naming, shaming and bragging. As Andresen and Gulbrandsen bring out in Chapter 8, public interest in climate change issues is dwindling in most Western societies, partly because the sophistication of the climate regime makes the contested issues rather technical and difficult to follow.

### *Facilitation*

Positive incentives, to a greater extent than sanctions, are currently the name of the compliance game in international environmental regimes whenever developing countries, economies in transition or other parties experience difficulties in complying with their commitments. This is clearly visible in the climate regime as well, and notable in the mandate of the Facilitative Branch of the Compliance Committee. Two subsidiary bodies under the FCCC, the Subsidiary Body for Scientific and Technological Advice and the Subsidiary Body for Implementation, will remain important for the regime's ability to bring life to this mandate.

The effectiveness of the Facilitative Branch will ultimately depend on the actual resources available for support activities. Experience from other environmental regimes suggests that funds for such supportive activities are likely to fall considerably short of the needs. The Global Environment Facility (GEF), set up to provide new and additional grant and concessional funding for environmental projects, including in the climate area, is the main international instrument for climate capacity enhancement and it is expected that some US$3 billion will be available for the 2002–2006 period, up from US$2.4 billion in the preceding fouryear period.[17] Environmental organizations seek to influence the operation of this instrument, especially with regard to transfer of technology to developing countries, by participating in the GEF's Ad Hoc Working Group on Global Warming and Energy (see Chapter 8 by Andresen and Gulbrandsen).

## *Enforcement: institutional and political constraints*

The more severe or 'hard' the enforcement measures available under a regime, the more important is procedural protection against their abuse. Meeting due process concerns is important for the acceptability of the compliance system. At the same time, one cost of extensive protection against unjustified claims of non-compliance is that excessive due process protections may cause the rejection of substantiated claims as well. For example, as a result of the voting procedures within the Compliance Committee explained above, two members from Annex I parties will in practice be able to prevent a question of non-compliance from moving forward for substantive consideration by the Enforcement Branch.

Members of the Enforcement Branch are to serve in their personal capacities. Some members might nevertheless consider the interests of their home countries when fulfilling their tasks. As shown by Hagem and Westskog in Chapter 5, the implications of such sideways glances at the national interest can be far-reaching. Especially if the party screened for non-compliance is a large economy, the suspension of its right to engage in transfers under Article 17 will influence price levels in the emissions trading market, and markets for fossil fuels and emission intensive products. In such cases, an Enforcement Branch member may be subject to pressure from its country of origin, since the decision of whether to sanction a non-compliant country could imply significant economic losses or gains, depending on the participation of its home country in these markets and its balance between imports and exports. Conceiveably, a party intending to be a net seller might expect Enforcement Branch members from parties that are buyers to vote against the suspension of its authorization to engage in transfers, since such a suspension would cause prices to rise markedly. This potential for influence may reduce the deterrent effect of the compliance system. It may also enhance the risk that over time the enforcement system might treat similar cases differently, undermining its own legitimacy (Franck, 1990).

## External means of enforcement

Given the strengths and limitations of the climate regime's internal compliance system, external means of enforcement might add significantly to the regime's overall deterrent effect. As argued by Hovi in Chapter 6, this is partly because external threats of punishment can imply more severe consequences for the non-compliant country. External threats might also, at least in some cases, carry more credibility than internal measures. At the same time, external pressure can supplement internal response options, thereby making internal means of compliance more effective than they might otherwise have been. In particular, external pressure might prevent a party from dropping out of the regime in response to the threat of internal sanctions.

On the other hand, external enforcement also entails certain potentially serious problems. Resort to external enforcement might undermine the regime's legitimacy, though this is likely to depend on the particular circumstances involved, and on the means and lawfulness of the external sanctions employed. The potency of climate-related trade measures, to take one prominent example, will be affected by how the climate regime interacts with the global rules administered by the World Trade Organization (WTO). At the core of this interplay, discussed by Stokke in Chapter 7, is the normative consistency of trade-related rules in the two regimes and any hierarchical relationship between them. The stronger clout of the WTO and its compulsory dispute settlement system suggest that issues involving competing claims would likely be referred to WTO bodies, which have so far been restrictive regarding exceptions to the general ban on discriminatory trade measures. Among WTO members, the findings of a dispute settlement body would presumably be affected by the status of the target under the Kyoto Protocol. A non-complier with Kyoto commitments would be more shielded than a non-party to Kyoto, because by joining the Kyoto regime a non-complier has exposed itself to internal and less trade-intrusive compliance measures that should be exhausted before resort to external compliance measures.

If external enforcement takes place within boundaries set by other international rules, and at the same time adds significantly to the deterrent effect of a regime's internal consequences, it is likely to enhance compliance without necessarily being detrimental to the regime's legitimacy. Thus although the best option is probably an internal system of enforcement that is both effective and fair, a second best may be external enforcement in accordance with other tools of international law.

## Compliance, NGOs and international governance

The climate regime illustrates the increasingly global character of environmental problems, and the need to resolve such problems through collective rather than individual action. States have, through the Kyoto Protocol, accepted far-reaching and binding commitments. While the climate regime is aimed at resolving an environmental problem, it will have significant effects on energy production and consumption, as well as our lifestyles. The regime's restrictions on sovereignty are thus not only important from an environmental point of view, but also economically and politically.

This importance is reflected in the mobilization of NGOs in the development of the compliance regime. As brought out in Chapters 1 and 8, environmental NGOs have consistently favoured a strong and transparent enforcement regime. The compliance system elaborated in the Marrakesh Accords conforms to a great degree to recommendations made by NGOs, although it is difficult to separate the distinct contributions of these organizations from contributions made by certain member states.

In Chapter 10, Wettestad points out that several multilateral environmental agreements share basic features in their systems for verification, review and response. However, as he also observes, there seems to be a tendency towards compliance systems with more teeth. While the Marrakesh Accords contain both a facilitative and an enforcement approach towards non-compliance, the enforcement aspect is clearly the most striking innovation. How important this innovation will be in practice depends on the ability of the provisions for hard enforcement to induce compliance. As pointed out by Barrett (2002; 2003), the Kyoto compliance regime suffers from a number of weaknesses that might potentially weaken the system. First, if a country that has been found by the Enforcement Branch to be in non-compliance in the first commitment period also fails to comply in the second period, it must presumably make up for the difference (plus 30 per cent) in the subsequent commitment period. However, this implies that the punishment might be forever delayed – an aspect of the system that obviously diminishes its deterrent effect. Second, an expectation of being found out of compliance in the first commitment period is likely to induce countries to hold out for a generous allowance in the upcoming negotiations for second period emission targets. This itself reduces the *de facto* punishment resulting from non-compliance in the first period.[18] Third, the Marrakesh Accords include no enforcement provisions to address the failure by a non-compliant country to comply with enforcement consequences. Hence, the compliance mechanism depends on the cooperation of the non-compliant party. Fourth, any party is entitled to withdraw from the Kyoto protocol upon giving 12 months' notice. It follows that a country that anticipates enforcement consequences might evade these consequences simply by withdrawing from the protocol.[19] Finally, it has been argued that the compliance mechanism is not legally binding, and can be made so only through an amendment which requires a three-quarters majority vote of the Meeting of the Parties; even if such an amendment is agreed, the compliance mechanism becomes binding only on those countries that choose to ratify this amendment.

The problem identified by Hagem and Westskog in Chapter 5 adds to this list. Even though the Marrakesh Accords state unambiguously that (all of) the hard consequences *shall* be imposed if a country is found not to be in compliance, it is not obvious that a non-compliant country actually will be punished. The reason is that 'determining whether or not a country is non-compliant requires judgment and discretion' (see Chapter 2 by Ulfstein and Werksman, and Rypdal et al, 2003). Moreover, if hard consequences are imposed on a non-compliant country, other countries that operate in related markets, especially for fossil fuels, will also suffer due to price effects. For this reason, some countries (such as Norway) might even suffer more if another country is punished than if they are punished themselves (Hovi and Kallbekken, 2004). Effects of this type are not only likely to be seen as unfair; they may also create incentives for strategic behaviour in the Enforcement Branch. If at least two members of the Enforcement Branch take these effects into consideration, a country that is technically in non-compliance might be able to avoid punishment.

Taken together, the above objections provide an important criticism of the compliance system. Yet it is conceivable that in practice, these weaknesses will not be as disruptive for its functioning as one might think. Proponents of the compliance system seem to believe that when hard consequences are imposed, the relevant country is likely to accept the punishment, the above objections notwithstanding. One reason why it might choose to do so is that it would otherwise risk jeopardizing the entire compliance system. Another reason is that, as emphasized by Hovi in Chapter 6 and Stokke in Chapter 7, other countries might use external pressure to convince a non-compliant party that it should accept the punishment.

Overall, it is therefore no easy task to predict how well Kyoto's compliance system is going to work in practice. The following chapters should give the reader a better understanding of the nuts and bolts of the compliance system and the factors that will determine its effectiveness.

## Notes

1 UNFCCC, Art 2. Developed countries, for this purpose, are listed in Annex I of the Convention. Annex II parties, a group comprised of those Annex I parties that are also Organization for Economic Cooperation and Development (OECD) members, have further commitments to provide financial and technological resources to enable developing countries to address their Convention commitments (UNFCCC, Arts 4(3), 4(4), 4(5)). In addition to their substantive commitments, developed countries also have more extensive reporting obligations than developing countries (UNFCCC, Art 12 (2)).

2 UNFCCC, Art 13.

3 UNFCCC COP Decision 10/CP.4.

4 The flexible mechanisms are: (i) emissions trading, (ii) joint implementation and (iii) the Clean Development Mechanism.

5 Marrakesh Accords, Decision 24/CP.7, Sec IV, Arts 4–6 (Facilitative Branch) and Sec V, Arts 4–6 (Enforcement Branch).

6 The usage of terms like 'verification' and 'review' varies significantly among authors; our usage is compatible with the definitions provided in UNEP's 2001 Guidelines on Compliance with and Enforcement of Multilateral Environmental Agreements.

7 Marrakesh Accords, Decision 24/CP.7, Sec XV.

8 Marrakesh Accords, Decision 24/CP.7, Sec II, Art 8–9.

9 Marrakesh Accords, Decision 24/CP.7, Sec XI, Arts 1 and 3; see Chapter 2 by Ulfstein and Werksman.

10 For leading contributions to the study of regime effectiveness, see Haas et al, 1993; Young, 1999; and Miles et al, 2002.

11 The 'logic of consequentiality' has been coined by March and Olsen (1989, pp23–4). Note that the enforcement school highlights the need for negative consequences of non-compliance, whereas the logic of consequentiality extends to capacity enhancement and rewards as well. See also Chapter 6 by Hovi.

12 Kyoto Protocol, Arts 5 (1)–(2) and 7–8.

13 Marrakesh Accords, Decision 24/CP.7, Secs IV and XIV; cfr Kyoto Protocol, Arts 5 and 7.

14 Marrakesh Accords, Decision 24/CP.7, Sec VIII, Arts 4 and 5.

15 Kyoto Protocol, Arts 3, 6, 12 and 17.

16 Kyoto Protocol, Art 3.5.

17 Stokke and Thommessen, 2003, p265.

18 According to the Marrakesh Accords, targets for the second commitment period should be agreed on well before the end of the first period and thus prior to any assertion of non-compliance. However, if past is prologue, one cannot take it for granted that this deadline will be met – and even if it is, states are likely to act strategically in light of at least rough expectations of whether they will be able to meet their Kyoto targets.

19 The option of withdrawal is of course available in most other international treaties.

## References

Barrett, S. (2002) 'Consensus Treaties', *Journal of Institutional and Theoretical Economics*, vol 158, no 4, 529–47

Barrett, S. (2003) *Environment and Statecraft*, Oxford University Press, Oxford

Brown Weiss, E. and Jacobsson, H. (1999) 'Getting Countries to Comply with International Agreements', *Environment*, vol 41, no 6, 16–20, 37–40.

Chayes, A. and Chayes, A. H. (1995) *The New Sovereignty: Compliance with Treaties in International Regulatory Regimes*, Harvard University Press, Cambridge, MA

Downs, G. W., Rocke, D. M. and Barsoom, P. N. (1996) 'Is the Good News about Compliance Good News about Cooperation?', *International Organization*, vol 50, no 3, 379–406

Franck, T. M. (1990) *The Power of Legitimacy Among Nations*, Oxford University Press, New York

Haas, P. M., Keohane, R. O. and Levy, M. A. (eds 1993) *Institutions for the Earth: Sources of Effective International Environmental Protection*, MIT Press, Cambridge, MA

Hagem, C., Kallbekken, S., Mæstad, O. and Westskog, H. (forthcoming 2004) 'Enforcing the Kyoto Protocol: Sanctions and Strategic Behavior,' *Energy Policy*

Hovi, J. and Kallbekken, S. (2004) 'The Price of Non-compliance with the Kyoto Protocol: The Remarkable Case of Norway', CICERO working paper 2004:7.

Jacob, TR (2002) 'Reflections on Bonn Climate Meetings', DuPont Senior Adviser, Global Affairs, memo, limited circulation by Tom.Jacob@USA.dupont.com, 17 June

Jacobsson, H. and Brown Weiss, E. (1998) *Engaging Countries: Strengthening Compliance with International Environmental Accords*, MIT Press, Cambridge, MA and London.

Loreti, C. P., Foster, S. A. and Obbagy, J. E. (2001) *An Overview of Greenhouse Gas Emissions Verification Issues*, Pew Center on Global Climate Change, Arlington, VA.

March, J. G. and Olsen, J. P. (1989) *Rediscovering Institutions: The Organizational Basis of Politics*, The Free Press/Macmillan, New York

Miles, E. L., Underdal, A., Andresen, S., Wettestad, J., Skjærseth, J. B. and Carlin, E. M. (2002) *Environmental Regime Effectiveness: Confronting Theory with Evidence*, MIT Press, Cambridge, MA

Rypdal, K., Stordal, F., Fuglestvedt, J. S. and Berntsen, T. (2003) 'Assessing Compliance with the Kyoto Protocol: Expert Reviews, Inverse Modeling or Both?', working paper 2003:07, CICERO, Oslo

Stokke, O. S. and Thommessen, Ø. B. (eds 2003) *Yearbook of International Co-operation on Environment and Development 2003/2004*, Earthscan, London

Werksman, J. (1996) 'Designing a Compliance System for the UN Framework Convention on Climate Change' in Cameron, J., Werksman, J. and Roderick, P. (eds) *Improving Compliance with International Environmental Law*, Earthscan, London, 85–112

Werksman, J. (1999) 'Greenhouse Gas Emissions Trading and the WTO', *Review of European Community and International Environmental Law*, vol 8, no 3, 1–14

Young, O. R. (ed 1999) The Effectiveness of International Environmental Regimes: Causal Connections and Behavioral Mechanisms, MIT Press, Cambridge, MA

# Part I

# The Kyoto Compliance Regime: Emergence and Design

Chapter 1

# The Negotiation of a Kyoto Compliance System

Jacob Werksman

## Introduction

At the seventh session of the Conference of the Parties (COP-7) to the UN Framework Convention on Climate Change (UNFCCC), the Convention's parties resolved many of the outstanding issues necessary to bring the Kyoto Protocol into operation. These agreements are contained in the report of COP-7 known as the Marrakesh Accords. One of the last pieces of the Kyoto machinery to be designed was the compliance system: a set of rules, procedures and institutions intended to 'facilitate, promote and enforce compliance' with the Protocol's commitments.[1] The compliance system represents over a decade of effort by climate change negotiators to tailor a mechanism to fit a regime whose features continuously shifted shape. The design choices that were made provide a crucial insight into the way in which delegations perceive the legal and political character of the their commitments under the Protocol.

This chapter explores, through a non-sequential, thematic negotiating history of the compliance-related elements of the Marrakesh Accords, the theoretical and political positions that underpin the Kyoto Protocol's compliance system. The starting point for the negotiations was Article 18 of the Protocol, which provides that:

> The Conference of the Parties serving as the meeting of the Parties to this Protocol shall, at its first session, approve appropriate and effective procedures and mechanisms to determine and to address cases of non-compliance with the provisions of this Protocol, including through the development of an indicative list of consequences, taking into account the cause, type, degree and frequency of non-compliance. Any procedures and mechanisms under this Article entailing binding consequences shall be adopted by means of an amendment to this Protocol.

Other than the procedural obligation to prepare a compliance system in time for adoption at the first session of the Conference of the Parties serving as the Meeting of the Parties (COP/MOP), the Article 18 mandate had only a limited impact on the outcome of the negotiations.[2] Instead, the negotiations were shaped by a series of design choices centred on the following themes:

1. The experience of the Convention parties in designing non-confrontational, facilitative approaches to non-compliance, and the shift in attitude of the majority of delegations towards the need for a 'strong' compliance system in light of the Kyoto Protocol's binding and quantified commitments and the introduction of market-based flexibility mechanisms into the climate regime.
2. The influence and attitudes of Annex I party delegations for and against the development of a strong compliance system, including domestic experiences in the US, regional approaches in the EU and the tension between 'due process' and 'automaticity'.
3. The influence and attitudes of developing country party delegations for and against the development of a strong compliance system, and tensions between progressive environmental objectives and concerns about the erosion of national sovereignty.
4. The role and contribution of non-governmental organizations (NGOs) in the development and testing of proposals during the negotiations, focusing on proposals for tough penalties and transparent procedures.
5. The impact of the fragmented structure of the Buenos Aires Programme of Action negotiations on the linkages between the compliance system, the Kyoto mechanisms and the reporting and review procedures.
6. The issue that both launched and ended the negotiations on a compliance system for the Kyoto Protocol: the legally binding character of the enforcement consequences.

After a brief linear history of the negotiations, and an overview of the compliance system, each of these themes is explored in more detail.

## Overview of the compliance system

The recent history of negotiation of the Kyoto compliance system begins with the adoption at COP-4 of the Buenos Aires Programme of Action, which set an agenda for the preparation for entry into force of the Protocol. It established a Joint Working Group (JWG) on Compliance with a mandate to 'develop procedures by which compliance with obligations under the Kyoto Protocol should be addressed'.[3] Developments since then can be traced through the submissions of parties to the JWG[4] and the reports and compilations made by the JWG co-chairs.[5] The penultimate stage of the negotiations was the Bonn Agreement, a political agreement on elements of the compliance system reached at COP-6, part 2, in Bonn.[6]

The Bonn Agreement resolved most of the main contentious issues that had divided delegations and that had contributed to the breakdown of negotiations – and the near collapse of the regime – at the first part of COP-6, in The Hague (Bodansky, 2001). Delegations reached COP-7 in Marrakesh, and the final stage of the negotiations, having agreed on the compliance system's objective, principles and scope, the consequences to be applied by the Enforcement Branch, the scope of an appeals procedure, and the size, composition and decision-making procedures of the Compliance Committee.[7] COP-7's compliance-related task was to translate these political agreements into procedures, mechanisms and institutions, and to decide on the legal form by which the compliance system would be adopted by the parties and come into force. The form of adoption for the compliance system and its implications for the legal character of its consequences proved to be two of the last and most difficult of the issues to be resolved.

As several chapters in this volume will illustrate, what emerged from these negotiations is a remarkable compliance system drawing on precedent from, and yet unique to, international law. Institutionally, the Kyoto system relies upon two functioning branches within a single Compliance Committee. The mandates of these branches are divided in accordance with the Protocol's commitments. While all commitments and all parties can be brought before the Committee's Facilitative Branch, the Enforcement Branch has exclusive jurisdiction over the specified, legally binding and target-related commitments of the Annex I (industrialized) countries. The technical compliance of the Annex I parties is reviewed annually by Expert Review Teams (ERTs), which are tasked with identifying any questions of implementation that arise from their reviews. Any party can also raise a question of implementation with regard to itself or with regard to any other party. The Bureau of the Compliance Committee will allocate the questions to one or the other branch, depending upon the commitment being questioned and the mandate of each branch. Each branch can then screen out questions it considers to be unfounded or *de minimis*. Following its deliberations, each branch has available to it 'compliance consequences' which it can apply to a party, with the Enforcement Branch wielding the harder 'enforcement consequences'.

The Enforcement Branch has the additional function, unique to the Kyoto Protocol, of policing the operation of the Protocol's flexibility mechanisms. These mechanisms allow parties with commitments to buy and sell emissions allowances and offsets internationally and to invest in emission reduction projects in developing countries in exchange for offsets. The effective operation of these mechanisms requires an accurate and transparent accounting system that can track the movement of credits from one party's national registry to another's. In order to remain eligible to participate in these mechanisms, the Annex I parties must meet the criteria for mechanism participation agreed in Marrakesh. The Enforcement Branch will review any questions raised regarding these criteria and has the authority to suspend and restore eligibility through a special expedited procedure.

The most distinguishing feature of the Kyoto Protocol compliance system is what delegations and observers refer to as its strength or toughness. This toughness is expressed through the enforcement consequences, and in the authority of a small body of experts to impose these consequences with only limited political oversight. While these consequences and the Enforcement Branch will be treated in detail in Chapter 2, they form a central aspect of the story of the negotiation of the compliance system as a whole. The toughness of these consequences can be divided into their prescriptiveness, their punitiveness and their legal character. Prescriptiveness describes the extent to which the system limits the discretion of the Enforcement Branch in determining what consequence would apply in a particular situation, by attaching an automatic, pre-determined response. Punitiveness describes the 'sting' or 'bite' of the consequence: in economic terms, the extent to which the consequence raises the costs of non-compliance above the costs of compliance.[8] The binding or non-binding legal character of the consequence describes the extent to which a party found in non-compliance is free, as a matter of law, to avoid the consequence imposed upon it.

## Political landscape of the negotiations

Before turning to the thematic analysis of the development of the compliance system, it is important to briefly sketch the political landscape surrounding the system's negotiation. The Protocol's political contours closely follow the shape of the Protocol's commitments. The first feature is the boundary between North and South; the industrialized countries contained in Annex I of the UNFCCC will have legally binding and quantified commitments under the Kyoto Protocol, while developing countries will not. Within Annex I, the emissions caps agreed for industrialized countries varied, and were expected to require very different levels of compliance effort. This is particularly the case along East–West boundaries, as many of the economies in transition from Eastern Europe and the former Soviet Union were assigned emission caps higher than what they are likely to need to meet their domestic requirements. But, as will be seen, attitudes towards non-compliance also varied among the wealthier industrialized countries, with sharp differences of view emerging on certain issues between the European Union,[9] and a number of groupings in the loosely-organized remainder of Annex I, known as the JUSCANZ and the Umbrella groups.[10]

Among the non-Annex I developing countries, interests in and attitudes towards compliance with Annex I's Kyoto targets varied as well. Generally, the developing countries negotiated as a bloc through their traditional caucus, the Group of 77 and China.[11] Their main common concern was to develop a compliance system that reflected the same 'common but differentiated' design as the Protocol's targets.[12] Larger developing countries, such as Brazil, India and China, were also concerned with the potential application of the compliance system to them, as the Protocol may evolve to include targets and timetables for non-Annex I countries. Smaller countries vulnerable to the potential impacts of climate change and broadly represented by the Alliance of Small Island States (AOSIS)[13] were primarily concerned about the effectiveness and enforcement powers of a compliance system. Countries dependent on the production and export of fossil fuels were concerned that a truly effective compliance system might move too many countries too quickly towards less carbon-intensive alternatives.[14]

One of the most significant political developments outside the climate change negotiations was the election to the presidency of the US, immediately following COP-6, part 1, of George W. Bush. A few months before COP-6, part 2, President Bush announced that his administration viewed the Kyoto Protocol as being 'fatally flawed in fundamental ways' and expressed its 'unwillingness to embrace' the treaty.[15] The US withdrawal from the Kyoto Protocol added momentum to the Bonn negotiations, particularly in areas where the US views had represented a stubborn and powerful minority. Some delegations were also likely driven by a desire to show the US that the international community was capable of finishing its work and of demonstrating its continued support for the Protocol without US involvement.

However, the withdrawal of the US from the negotiations did not have a major impact on the outcome of the compliance system. As will be seen, since Kyoto the US had consistently supported a strong compliance system and, with one important exception (the composition of the Enforcement Branch), many of the elements adopted at Marrakesh had either derived from US submissions or were supported by the US delegation at some point in the negotiations.[16]

While much of the final design of the compliance system can be explained by contributions and trade-offs between political groupings, significant credit for the shape of the final package must be given to various individuals who chaired or steered the process.[17] The topic had the potential to become politically-charged, but up until the last few sessions of the JWG, the discussions maintained an analytical, at times almost academic atmosphere. The JWG invited presentations from academic institutions, NGOs and the secretariats of other institutions with non-compliance and dispute settlement mechanisms. The JWG was also aided by the consistent participation of a significant core group of individual delegates who provided institutional memory and a momentum behind the process. This atmosphere of trust allowed the JWG co-chairs to make a number of significant consolidations and advances in the development of the text.[18]

## From facilitation to enforcement: Conventional wisdom revisited

Assessing the extent to which the Kyoto Protocol's compliance system reflects a major shift in thinking among delegations requires a quick look back at previous efforts to develop compliance systems for the climate change regime. The compliance system is the third to have been attempted by climate change negotiators over the past decade. The first effort, made during the negotiation of the Convention, was abandoned in favour of an 'enabling provision' in Article 13 of the UNFCCC, which calls upon the Convention parties to 'consider the establishment of a multilateral consultative process ... for the resolution of questions regarding the implementation of the Convention'. As it became apparent during the negotiations of the FCCC that the treaty would not result in specified, legally binding emissions reduction commitments, delegations determined that a compliance system could await the development of more specific treaty commitments.

The second effort followed the entry into force of the Convention and was based on the Article 13 mandate. This time, negotiators built a procedure in its entirety, but the parties to the Convention have yet to bring it into operation (Szell, 1995). As part of a negotiating process that ran in parallel to the negotiation of the Kyoto Protocol, climate negotiators designed a multilateral consultative process (MCP), and agreed on all elements but the size and composition of the institution that would run the MCP.[19] The main characteristics of this procedure reflect what, from a theoretical point of view, has been described as a 'managerial approach' to non-compliance (Downs et al, 2000; Danish, 1997).[20] The design of the MCP was heavily influenced by the experience of the design and initial operation of a non-compliance system for the Montreal Protocol on Ozone Substances that Deplete the Ozone Layer (Koskenniemi, 1992; Victor, 1998; Werksman, 1996; see also Chapter 10 by Wettestad). A multilateral committee, comprised of experts nominated by the parties, offers technical advice and financial assistance to parties facing compliance difficulties in an effort to head off potential non-compliance. Liberal standing provisions, non-confrontational procedures and soft consequences invite parties experiencing difficulties to a constructively engage with the system. The MCP's objectives are to provide advice on assistance to parties to overcome difficulties encountered in their implementation of the Convention, to promote understanding of the Convention, and to prevent disputes from arising. Its mandate is limited to providing advice and making recommendations. Like the Montreal Protocol compliance system, recommendations made by the climate change regime's MCP would have to be approved by the parties as a whole.

The MCP was designed to apply to the Convention and its vague obligations. However, the MCP was being negotiated in parallel to the Protocol, and delegations could have opted simply to apply this emerging managerial system to the tougher commitments emerging from the Protocol.[21] It is important to recall that advocates of the managerial approach to non-compliance have not tied their theories supporting the effectiveness of a soft approach to soft commitments. On the contrary, soft approaches to compliance are thought to make tougher targets more politically acceptable. Non-confrontational approaches are indeed made necessary by difficult commitments, and favour the pragmatism of engagement over the futility of enforcement against notoriously resistant sovereign states. As has been indicated, the model precedent for the managerial approach was the Montreal Protocol non-compliance procedure, which contains highly specific, legally binding commitments.

Nonetheless, as the Protocol took shape, delegations' attitudes towards non-compliance procedures shifted away from the managerial approaches represented by the MCP and towards tougher enforcement procedures. Three basic reasons can be provided for this shift:

1 Competitiveness concerns in the context of ambitious targets demand a means for identifying and discouraging parties from free-riding by making the costs of non-compliance higher than the costs of compliance.
2 The introduction of market-based instruments carries with it the need to assure traders in carbon allowances and offsets that the benefits of their bargains will be backed by a rule-based response.
3 The legally binding character of the Kyoto Protocol's targets demands a compulsory and binding means of enforcing them. (Werksman, 1999)

The first and second reasons provide the clearest rationales for departing from the managerial approach. The managerial approach is designed to respond to parties that wish to comply, but lack the financial and technical means of doing so. The Montreal Protocol system has worked well, in large part because it has focused on solving the compliance problems of those countries that are eligible to receive financial and technical assistance from international financial mechanisms such as the Multilateral Fund of the Montreal Protocol and the Global Environment Facility (GEF). The financial and political costs of compliance to these countries are lowered through the transfer of technology and financial resources.

In contrast to the Montreal Protocol, many perceive the Kyoto Protocol's commitments as imposing serious economic and political costs on industrialized countries. These countries are donors rather than recipients of development assistance. They also tend to be drivers of technological innovation and therefore the home of innovative industrial patents. For these reasons, traditional compliance incentives such as transfers of financial and technological resources make little sense as a response to the parties most likely to exceed their Kyoto caps. Thus, at the outset of negotiations, delegations from North and South were calling for a 'strong and effective' compliance system that would 'provide parties with certainty and confidence'.[22]

The second reason also illustrates the central role that Kyoto's market-based flexibility mechanisms[23] played in the design of the compliance procedure. These mechanisms will allow Annex I countries and companies to meet their obligations by investing in emissions reduction opportunities in other countries, wherever the costs of compliance are lowest. By lowering the cost of compliance they provide a crucial safety valve for countries struggling to comply. However, markets in these offsets depend upon the regulatory incentives created by a credible compliance system. Offsets and allowances only take on a marketable value when they are in demand by regulators as part of a strong compliance system. This close design link between the strength of the compliance procedure and the effective operation of the Protocol's market-based mechanism was key to building a consensus around a strong compliance system. Even those Annex I delegations that remained ambivalent with regard to tough enforcement consequences found it difficult to square a soft approach with their unshakable enthusiasm for the Protocol's market mechanisms.[24]

The third reason is backed by a somewhat compelling logic: as a regime's targets are strengthened, so should the procedures and mechanisms to enforce it. This view is not, however, always reflected in the design of multilateral environmental agreements, and, as has been indicated, this logic is not always supported by the academic literature (see the Introduction).

The essence of the managerial approach does, however, remain within the Facilitative Branch of the Marrakesh procedure, which shares many characteristics with both the Montreal Protocol system and the Convention's MCP. To a lesser extent, it has also left its mark on the Enforcement Branch. Although the Enforcement Branch has been authorized to impose substantial sanctions, and has a number of quasi-judicial/administrative tribunal-like features, the branch continues to operate in a relatively multilateral, preventative and non-confrontational fashion. For example, a party that initiates the enforcement process does not take on the role of complainant or prosecutor, but merely triggers the process, which is then left to the Compliance Committee and its branches.

Kyoto's compliance system does reflect a shift in political attitude that may undermine the theoretical underpinning of the managerial approach. It would appear that many of the assumptions of the managerial approach are no longer considered by policy-makers to be of general application. After Kyoto, facilitation appears to be considered appropriate, but only to regimes or those aspects of regimes where non-compliance is likely to be attributable to a lack of technical or financial resources. However, the multilateral nature of the Kyoto system suggests that theorists and practitioners continue to agree on the importance of a process of engagement focused on preventing rather than punishing non-compliance.

## Annex I 'one-upmanship'

The main focus of negotiations was on the development of enforcement procedures rather than on facilitative procedures, although developing country delegations played an important part in shaping the Kyoto compliance system. The central negotiating dynamic developed among those developed country parties to which tougher enforcement procedures would apply. As countries with commitments, Annex I parties arguably had the deepest interests in the design of the compliance system. Although their negotiating positions were more diverse than those of developing countries, the majority of Annex I countries shared a basic set of objectives throughout most of the negotiating process. The US, Canada, New Zealand, Switzerland and the EU shared a common commitment to designing a strong and effective compliance system, characterized by an enforcement function.

The enthusiasm of these countries for a strong compliance system in the abstract seemed to feed on itself, leading to a kind of virtuous competition. However, when the strength of the system was expressed in terms of concrete proposals, strong theoretical and cultural divides emerged. US proposals for a strong compliance system derived primarily from its experience with its own Clean Air Act and related emissions trading schemes designed to regulate acid rain emissions from large power plants.[25] The main lesson the US drew from this experience was that a successful regime is based on clearly defined penalties enforced against emitters in a highly predictable and automatic manner. The market depends upon a predictability specifically derived from the automatic nature of these penalties, which impose a cost on the emitter that is well above the market price of permits. The penalties are, in practice, never enforced, as emitters use the market mechanisms to remain within their targets, driven by the desire to avoid the higher cost of non-compliance. An obvious assumption underlying this approach is that the majority of the market players in the Kyoto regime would be private companies that could either buy themselves into compliance, or reduce their emissions as they are forced out of business. On the basis of this experience, US proposals, backed by other supporters of the market mechanisms, were aimed at limiting the discretion of the Enforcement Branch to a technical exercise of comparing emissions to targets. It also led to a reluctance to include the suspension of eligibility to participate in the mechanisms – the supposed route towards compliance – as a consequence of non-compliance. As is discussed in more detail in Chapter 2, under the US approach, the automaticity of the procedure helps to deliver due process by limiting the potential for political interference, and ensuring that all parties in the same circumstances receive the same treatment.

By contrast, European negotiators approached the Kyoto compliance system with what appeared to be a less mechanical view of enforcement. EU positions emphasized the need for domestic action, and sought to cap the extent to which any one party could rely on emissions reductions generated outside its jurisdiction to meet its target (see also Chapter 10 by Wettestad). Once use or access to the flexibility mechanisms is capped in some way, an enforcement system must begin to contemplate how best to respond to a party that has failed to meet its domestic obligations. Such a system may need to take a more nuanced approach to non-compliance, looking at each party on a case-by-case basis and constructing domestic compliance action plans that touch upon the individual circumstances of the party concerned. It could be argued that European negotiators' membership in a supra-national community with central enforcement powers influenced their view of the political acceptability of this more intrusive approach. Some have also suggested that the EU's preference for a case-by-case approach resulted from its inability to agree early in the negotiations on what pre-defined penalties would be appropriate. Under the European approach, due process is guaranteed by a case-by-case approach which involves hearings and the considered judgement of experts, rather than automaticity.

Within the Annex I grouping, but on the fringe of the central debate, were delegations from Japan, Australia and the Russian Federation, which rejected the need for a heavy-handed enforcement system. Until the final stages of the negotiations, they remained unconvinced that the Kyoto Protocol's targets and mechanisms were sufficiently different in their character to justify a revolutionary new approach to enforcement. Good faith and non-binding consequences should, according to these delegations, suffice under international law.

The choice between a purely facilitative and an enforcement approach also depends, in part, on the perceptions of the target non-complier. The negotiating stances of many of these delegations seemed to be predicated on the assumption that it would be a country other than its own that would find itself facing the Enforcement Branch. Despite domestic assessments of the likely high economic and political costs of its target, the US delegation consistently sent the message that once the US commits to a target, the robustness of its domestic legal and regulatory system, in combination with the market mechanisms, would deliver the results.[26] US scepticism seemed to focus on the risk of a 'climate rogue' that might, for example, oversell and then exceed its emissions allowance. The US delegation also pressed hard for close scrutiny of the EU's internal arrangements under its 'Article 4 bubble'.[27] The Europeans seemed equally confident that their compliance would not be challenged. They reserved most of their scepticism for the operation of the Kyoto mechanisms, out of concern, perhaps, that the US, having inspired and designed these instruments, would find some means of taking advantage. Advocating a strong approach also implicitly assumes that the non-compliance of others will either be successfully deterred, or at the very least it will be contained to relatively few parties. Widespread non-compliance confronted with tough penalties would likely generate either defections from the regime, or a renegotiation of its rules.

From this perspective, the Japanese, Australian and Russian minority positions take on a degree of reasonableness, in that they were willing to imagine the possibility of their own non-compliance. The Kyoto targets are both difficult and arbitrary, and in the months and years that have followed their adoption, many countries have begun to reassess how challenging these targets may be to reach. Japan and Australia may have taken less comfort in the safety valve offered by the Protocol's market mechanisms, as many predicted that the US and US companies would corner the market in cheap offsets.

Ironically, Russia, with its generous target, was (for many observers) the Enforcement Branch's most likely client. Although on its own Russia would have more emissions allowances that it could use during the commitment period, these additional tonnes, known as 'hot air', have attracted substantial controversy. During the negotiation, rumours circulated that the Russian government had already promised its tonnes in bilateral deals with other sovereigns, and that Gazprom, Russia's partially privatized gas monopoly, had promised the same tonnes to others.[28] The weakness of Russia's domestic institutions raised additional concerns that reductions would not be measured reliably, and that the ERTs and the Enforcement Branch would be forced to reveal these weaknesses.

Although the US played only a background role in the final stages of the negotiation of compliance system, the main elements of its vision for a strong and automatic enforcement system remain intact. European discretion will, however, be introduced through the inclusion in the enforcement consequences of a Compliance Action Plan for each Annex I party found to have exceeded its cap. The plan, which will be reviewed and assessed by the Enforcement Branch, will include an analysis of the causes of non-compliance and describe the actions the party intends to implement to meet its target, 'giving priority to domestic policies and measures'. In the end, Australia, Japan, Canada and Russia accepted the compliance system's tough and automatic consequences, but managed to block consensus on the legally binding character of those consequences. The implications of this result are discussed below.

## Developing countries and the burdens of sovereignty

Given the potential divergence of interests, developing country negotiators, through the Group of 77, remained remarkably united and consistent in their participation in the design of the compliance system. Although individual delegations contributed specific proposals, the Group was strongest in its insistence on two design elements: a strong enforcement system, and one applicable exclusively to Annex I parties.[29]

While the rationale for their general support for the second element is clear, explaining the first element of the G-77 position is not straightforward. As has been described, although developing countries will not have specific and binding commitments under the Protocol and will not be subject to its jurisdiction, they do have a longstanding tradition of defence against international organizations making inroads into sovereignty. If one compares, for example, China, Samoa and Saudi Arabia, these G-77 members have very different national interests with regard to the long-term development and effectiveness of the climate regime. Nevertheless, the first major submission made by South Africa on the Group's behalf contains clear support for prescriptive, punitive and legally binding consequences.[30] As the negotiations reached their final round, developing country positions became more nuanced and fragmented, and in the last round seven developing countries made submissions without a common group paper.[31] However, none of the submissions strayed from the basic group position in support of a strong compliance system. On the contrary, a number of developing country delegations supported the inclusion of even tougher sanctions than those contemplated by Annex I countries, including the imposition of financial penalties.[32]

It would be possible to conclude that developing countries were comfortable in pressing for a tough compliance system in anticipation that even developing countries with significant emissions could avoid commitments indefinitely. Under this theory, developing country negotiators would be comfortable in designing a tough system in anticipation that it would never apply to them. The G-77 did, indeed, insist that the mandate of the Enforcement Branch clearly limited its jurisdiction to parties included in Annex I. The struggle to get developing countries to undertake quantified and binding commitments under the climate regime will be a long and difficult one. However, the design of the compliance system does not support the conclusion that once that threshold has been crossed, the jurisdiction of the Enforcement Branch will remain exclusive to industrialized countries. The distinctions in jurisdiction are based on the status of the party under the Protocol (whether it is included in Annex I, and whether it has a commitment under, for example, Article 3.1 of the Protocol), and not whether it is considered developed or developing.

A number of developing countries sought to build in procedural protections against the power of the Compliance Committee, efforts which could be explained as anticipating a time in which the same tough consequences that applied to industrialized countries might apply to themselves. Brazil, for example, consistently raised the need to provide the compliance system with a degree of political oversight, to prevent 'mere technicians' from imposing enforcement consequences on sovereign states.[33] Brazil's proposal, which received some support from other G-77 members,[34] was that the COP/MOP should, through a subcommittee of party representatives, provide a kind of political filter at both ends of the compliance process. For Brazil, both the initial questions of implementation and the ultimate findings of the Compliance Committee would need to be vetted by the COP/MOP subcommittee. However, even this proposal would have placed a relatively mild constraint on the power of the Compliance Committee. In Brazil's last iteration of the role of the COP/MOP subcommittee, that subcommittee would only have been able to overturn a decision of the Compliance Committee by consensus (the so-called reverse consensus rule).[35] Annex I countries overwhelmingly resisted a prominent role for the COP/MOP in regulating the outcome of the Compliance Committee. They saw the potential politicization of the process by a subcommittee of parties, or by the COP/MOP as a whole (where developing country diplomats were likely to dominate), as a greater threat to their sovereignty and to the integrity of the process, than the prominent role proposed for the Compliance Committee's technicians and experts. As the Annex I countries' sovereignty was more immediately at stake in the process, it was difficult to build support for Brazil's proposal.

Another potential source of resistance to a strong compliance system might have been expected from oil-exporting developing countries. OPEC (Organization of Petroleum-Exporting Countries) nations have been frequently accused of playing an obstructionist role in the negotiations of a climate regime. They have, however, always played a subtle game, often resisting but never fully blocking progress.[36] In light of the wide support, within the rest of the G-77, for a strong compliance system, the approach of OPEC delegations was to broaden the mandate of the Enforcement Branch, rather than to narrow its powers. In particular, Saudi Arabia sought to build support within the G-77 for the application of tough compliance consequences to the Protocol's commitments related to the potential impacts of Annex I parties' response measures on countries particularly dependent on the production of fossil fuels.[37] Such a strategy appeared to be aimed either at strengthening the legal character of these rather loosely-worded commitments, or at weakening the design of the enforcement system by linking it to vague and unenforceable provisions. While the proposal gained some support in early G-77 positions, it eventually fell away to the more compelling logic of a compliance system focused on clear and precise commitments.

A further, specific protection that developing countries sought in the design of the compliance system was their insistence on equitable geographic representation within all institutional aspects of the system, including in the Enforcement Branch, even though this branch would not review developing country commitments. The G-77 pressed for a membership formula based on the UN's five traditional regional groups, a formula that would ensure a majority of developing country slots. While the EU signalled flexibility towards this approach, the Umbrella Group steadfastly resisted the notion of a majority of developing country appointees sitting in judgement on Annex I performance. As is discussed in greater detail in Chapter 2, this proved to be a very contentious part of the negotiations and contributed to the breakdown of at COP-6, part 1, in The Hague.

When delegations were reconvened for part 2 of COP-6, in Bonn, the Chairman of the negotiations (Jan Pronk of the Netherlands) held a high-level all-night session aimed at resolving the remaining differences (Lefebere, 2002). The EU's gestures of accommodation towards the G-77 exposed the Umbrella group members as the ones who were holding out, and Pronk deployed a classic technique for reaching compromise: he requested the Umbrella countries to prioritize their concerns. By this point, the Umbrella group's numbers and strength had been severely diminished by the defection of Norway, Iceland and New Zealand,[38] and by the now symbolic participation of the US (which had renounced the Protocol). The group's remaining members set their priorities as follows:

1 Remove the possibility of financial penalties as an enforcement consequence.
2 Clarify that the mode of adoption of the compliance system would not lead to legally binding consequences.[39]
3 Provide for exclusive or majority participation of Annex I parties in the composition of the Enforcement Branch.

The G-77, which had never seriously believed that financial penalties would be part of the final package,[40] converged with the EU on the Chairman's strategy, and isolated the rump of the Umbrella group. The concept of financial penalties was removed from consideration and, in exchange, a modified version of regional geographical representation was adopted for the composition of the Compliance Committee, including the Enforcement Branch. This would give non-Annex I (developing countries) a majority of nominees on the branch.[41]

Although the balance of membership within the Enforcement Branch will remain a sore point for some Annex I countries, if the branch operates in a professional manner the strong participation of non-Annex I countries could provide an important point of leverage for the fuller participation of developing countries in binding commitments and in the enforcement procedures at a later stage in the development of the Kyoto regime. As is discussed below, the mode of adoption of the compliance system, and the linked issue of the legal character of the enforcement consequences, was left open for discussion at COP-7.

## NGOs: Experts and observers

A small number of persistent NGOs closely tracked the development of the Kyoto compliance system. Their strategies and influence are closely analysed in Chapter 8. NGO experts directly advised and shaped the positions of national delegations. They were allowed to present views and options at a number of workshops that were key to the development of the compliance system. Their consistent support for a strong enforcement mechanism, and their ability to articulate how such a system could work in practice, helped to maintain the focus of the negotiations on the need for an effective Enforcement Branch.

The NGO community in the final stages of the negotiations sought to ensure that this role could continue once the compliance system became operational. While there was wide support for the compliance system to be open and transparent, a few delegations, in particular the Russian Federation, resisted. Russia was believed by many NGOs to be the party most likely to have difficulty in supplying credible data on its compliance. Russia may have felt some concern that open access for NGOs would heighten scrutiny of its performance.

As described in Chapter 8, NGOs were generally successful in ensuring that Enforcement Branch deliberations, hearings and documentation would be open to contributions from NGOs and the presence of NGO observers, unless in special circumstances the Branch decided otherwise. These are essentially the same apparently liberal rules that apply to the Convention and Protocol bodies as a whole. However, the negotiations also tell a cautionary tale that shows liberal rules do not always lead to liberal practice. Ironically, under these very rules, NGOs were excluded from observing most of the final stages of the negotiations of the compliance procedure on the grounds that the issues being discussed were too sensitive.

## Loose ends from a fragmented negotiating process

The highly complex nature of the Kyoto Protocol, made necessary by the complex nature of the problem to which it responds, presented serious practical challenges to the negotiators. The compliance system sits at the heart of the operation of the Kyoto flexibility mechanisms and of the reporting and review procedures. However, during the negotiations, each of these issues was assigned to a separate working group. At several stages in the negotiations, delegations and the chairs of the various groups sought to ensure that cross-cutting and interlinking issues such as the links between the mechanisms and compliance were either clearly assigned to one negotiating group, or that they were taken in up in parallel in a conscious manner.

One specific issue that will be critical to the compliance system was negotiated in its entirety outside the JWG by another group: the definition of non-compliance. Coming to an understanding of how exactly non-compliance or potential non-compliance will be identified is important at both ends of the operation of the compliance system. At the point of initiating the procedure, the Compliance Committee must determine whether or not a question of implementation that has been raised is *de minimis* in nature. At the end of the procedure, a declaration of non-compliance may also need to take into account margins for error. Although these issues are fundamental legal concerns, they were never taken up by the JWG's legal experts. The implications of the difficulties in defining non-compliance are taken up in Chapter 2.

A further controversial issue, and the last to be resolved by negotiators in Marrakesh, was the relationship between the compliance system and the rules that would determine eligibility to participate in the Kyoto mechanisms. As has been described, the Enforcement Branch will have the authority to review questions of implementation raised with regard to an Annex I party's eligibility to participate in the Protocol's flexibility mechanisms, and the authority to suspend and reinstate that eligibility. In order to cement this essential aspect of the compliance/flexibility relationship, a number of delegations sought to introduce, in the negotiating group on mechanisms and in the JWG on compliance, a solid link between the two systems. In the mechanisms group, these delegations pressed for the express inclusion in the mechanism 'eligibility criteria' the commitment by each party to subject itself to the Enforcement Branch of the compliance system. The purpose of such a link would be twofold: to emphasize that the integrity of the flexibility mechanisms depended on the integrity of the compliance system, and to provide a strong incentive for all Annex I parties to adopt, adhere to and, potentially, ratify the compliance system, depending on what legal form the system would take. The issue of linkage was thus further complicated by the fact that the negotiators had not yet agreed on the legal character of the consequences and the means by which the compliance system would be adopted and become effective. While most delegations were comfortable having the Enforcement Branch review mechanism eligibility, some were adamantly opposed to making ratification of a compliance system with binding consequences a pre-condition for mechanism eligibility.

In the end, the link was left strongly implicit. The rules on the mechanisms agreed in Marrakesh provide that the:

> eligibility to participate in the mechanisms by a Party included in Annex I shall be dependent on its compliance with methodological and reporting requirements [and that] [o]versight of this provision will be provided by the Enforcement Branch of the compliance committee, in accordance with the procedures and mechanisms relating to compliance as contained in decision 24/CP.7, assuming approval of such procedures and mechanisms by the Conference of the Parties serving as the meeting of the Parties to the Kyoto Protocol in decision form in addition to any amendment entailing legally binding consequences, noting that it is the prerogative of the Conference of the Parties serving as the meeting of the Parties to the Kyoto Protocol to decide on the legal form of the procedures and mechanisms relating to compliance.[42]

Despite its convoluted caveat about eventual decisions by the COP/MOP, the decision indicates that eligibility for the mechanisms must be reviewed and that the review will be made by the Enforcement Branch. No manner of alternative is provided for an Annex I party to participate in the mechanisms without having subjected itself to the jurisdiction of the Enforcement Branch. This suggests that if COP/MOP-1 decides to adopt a compliance system with legally binding consequences that are subject to ratification, mechanism eligibility will be tied to participation in that compliance system and, by implication, ratification of the binding consequences.

## Legally binding consequences: The alpha and omega of the negotiations

As has been seen, the issue of the legal character of any binding consequences that would result from the Kyoto compliance system played a central role in the negotiations. At the outset, the mandate contained in Article 18 of the Protocol set an expectation and an ambition that binding consequences would form part of an effective compliance system. However, Article 18's links between such consequences and the need for an amendment to the Protocol complicated the situation and thus constrained those pressing for a clear legal outcome.

Article 18 of the Protocol requires that '[a]ny procedures and mechanisms under this Article entailing binding consequences shall be adopted by means of an amendment to this Protocol'. Delegations on all sides of the debate recognized that the regular amendment procedures under Article 20 of the Protocol would raise problems with regard to uncertainty, timing and the general application of the amendment to all parties. Waiting for the COP/MOP to adopt an amendment, and for all parties to ratify it, could leave a long period during which at least some parties would not be bound by the agreed consequences. It also left open the question of the status of those parties that wished to participate in the Protocol but were unwilling or politically unable to ratify a compliance amendment.

There were various attempts to finesse or exploit these constraints in order to either secure or prevent agreement on legally binding consequences. Those parties that were in support of legally binding consequences, including the EU, the G-77 and the US, recognized the difficulties of finding a legally sound manner of providing them with a legal basis that was consistent with Article 18 and that would be universally applicable. These delegations were intent on finding a creative solution. Those that did not want binding consequences, including Japan, Australia and Canada, sought to stick to the letter of Article 18 and to cite the complexities of amendment as another good reason to opt for a softer, politically binding solution.

As delegations approached COP-6, part 1 in The Hague, four overlapping options were offered by the co-chairs.[43] Option 1 would have recommended the adoption of the compliance system by decision, without reference to its legal character. Option 2 would have asserted the authority of COP/MOP-1 to adopt the compliance system as a legally binding decision, without reference to the need for any additional procedure to give it legal force. The first of these options was considered by most as weak, the second as having little or no legal basis.

Option 3 would have required the adoption at COP-6 of a legal instrument containing the entirety of the procedure, which would be available for ratification by parties in conjunction with the ratification of the Protocol. This proposal assumes that those countries that have already adopted, signed and even ratified the Protocol are prepared effectively to amend it prior to its entry into force, and that all delegations undertake to ratify both instruments as a package. The proposal has the advantage of providing immediate certainty as to agreement on the compliance package, but relies heavily on political will to ensure that all ratify both the protocol and the package.

Option 4 would recommend to a future COP/MOP that it adopt, as part of the legal instrument adopting the targets for the *second* commitment period still to be negotiated, the binding consequences applicable to the first commitment period. This proposal assumes that all countries committed to the long-term success of the climate regime will be prepared to adopt and ratify their new second period commitments before the end of the first commitment period. The proposal would provide a sound legal basis for these countries, but would provide a long time delay and uncertainty between Marrakesh and the eventual adoption by the COP/MOP of the final package.

In Bonn, at COP-6, part 2, the ministers reached consensus on the need for tough consequences and agreed terms that described the application of these consequences in clear and mandatory terms. They failed, however, to agree on how to bring the compliance system into force. The rump of the Umbrella group sought to interpret the failure of ministers to agree as to *how* the compliance procedure would come into force as a failure to agree *whether* the consequences resulting from the procedure were intended to be binding. This would have suggested, contrary to what many saw as the plain meaning of the Bonn Agreement, that delegations had viewed the Enforcement Branch consequences as part of an optional or 'menu' approach. This tactic nearly led to the breakdown of discussions in Bonn. A new approach emerged at COP-7, which was to suggest that the Bonn Agreement had not pre-judged whether COP/MOP-1 would decide to adopt compliance procedures entailing binding consequences. The G-77 and the EU felt that the choice of ministers in the Bonn Agreement to describe the proposed consequences in clear and mandatory terms, including through the use of the word 'shall', in fact did pre-judge the binding nature of those consequences.

The Marrakesh Accords do not resolve this fundamental dispute about the legal form of the compliance procedure. The issue was deferred, in neutral terms, to the first COP/MOP, by '*[n]oting* that it is the prerogative of the Conference of the Parties serving as the Meeting of the Parties to the Kyoto Protocol to decide on the legal form of the procedures and mechanisms relating to compliance'.[44] The implications of having left this aspect of the regime open are discussed in detail in Chapter 2.

## Concluding observations

As with many other aspects of the Marrakesh Accords, the Kyoto compliance system has been put on ice until the first session of the COP/MOP, which will take place within a year of the Protocol's entry into force. Its most controversial aspects – the enforcement consequences applicable to an Annex I party in breach of its emissions cap – will not be fully tested until after the end of the first commitment period, more than a decade away. It is extremely difficult, in a world of shifting politics and economies, to predict how well the compliance system will stand up to the political will of that moment.

Before then, the design of the compliance system and its theoretical and political underpinnings should be tested regularly by domestic legislatures as they decide whether to ratify the Protocol. The reason for completing the Marrakesh Accords before COP/MOP-1 was to give these legislatures the opportunity to have a fuller picture of how the regime will operate. Although many of the issues that were raised as the compliance system was being designed can, in theory, be reopened at COP/MOP-1, the issue of legal form of adoption remains an open and controversial issue. The process of ratification should build an even stronger and better informed political consensus around the Protocol and the Marrakesh Accords. If it does, the compliance system, backed by this deeper and stronger political will, should remain intact, and the legal form of its adoption will become less relevant to its success.

## Notes

1 The text of the completed Kyoto compliance system is contained in the Annex to decision 24/CP.7, FCCC/CP/2001/13/Add.3, taken by COP-7. COP-7 recommends that the first session of the Conference of the Parties serving as the Meeting of the Parties to the Protocol (COP/MOP) adopt the Annex.

2 The Joint Working Group (JWG)'s mandate, adopted as part of the Buenos Aires Programme of Action, did not mention Art 18, but instead referred to all compliance-related elements of the Protocol. This was in part to ensure that the JWG took into account compliance issues outside Art 18, such as rules related to the Kyoto mechanisms; and in a part an effort by some delegations to avoid the procedural dictates of the last sentence of Art 18, which links binding consequences to the Protocol's troublesome amendment procedures. Importantly, the Kyoto compliance system's indicative list of consequences is applicable only to its facilitative functions. Its enforcement functions lead to a series of compulsory consequences. FCCC/CP/1998/16/Add.1, Annex II [hereinafter Buenos Aires Programme of Action].

3 Buenos Aires Programme of Action.

4 FCCC/SB/1999/MISC.4 and Add 1, 2, 3, 29 April 1999 [hereinafter April 1999 submissions]; FCCC/SB/1999/MISC.12, Add 1-2, 22 September 1999 [hereinafter September 1999 submissions]; FCCC/SB/2000/MISC.2 [hereinafter 2000 submissions].

5 FCCC/SBI/2000/17, Report of the JWG, Report on SBI 13 part II; FCCC/SB/2000/10, Report of the JWG, Report on SBI 13 part I, Annex III; FCCC/SB/2000/5, Report of the JWG, Report on SBI 12, Annex III; FCCC/SB/2000/1, Elements of a Compliance System for the Kyoto Protocol. Note by the JWG co-chairmen, FCCC/SBI/1999/8 Report of the JWG, Report on SBI 10, Annex II.

6 FCCC/CP/2001/5, pp48–49, Report of COP-6, Part 2. 25 September 2001 [hereinafter the Bonn Agreement].

7 See the Bonn Agreement.

8 In the Bonn Agreement, delegations referred to the purpose of enforcement consequences as 'aimed at the restoration of non-compliance … ensur[ing] environmental integrity, and provid[ing] for an incentive to comply.' This language sought to avoid the implication that the system would impose sanctions. It also suggests that the object was not to punish non-

compliance but rather to ensure that the environmental benefits lost by delay were made up for by an increased effort.

9 The Member States of the EU were led by the delegation representing the presidency. Major submissions on compliance were made by Germany (April 1999 submissions), Finland (September 1999 submissions) and Portugal (2000 submissions). The last two of these were made also on the behalf of certain accession states, including Bulgaria, the Czech Republic and Cyprus.

10 JUSCANZ is roughly made up of Japan, US, Switzerland, Canada, Australia, Norway and New Zealand. With the exception of Switzerland, the JUSCANZ members had formed the core of the Umbrella group, which also included Russia. Switzerland joined with Mexico and South Korea (the only other OECD members that are not members of the EU or JUSCANZ) to form the Environmental Integrity Group. As will be seen, disagreements over compliance issues led to the further fracturing of the Umbrella group.

11 The G-77's coordinator and spokesperson for compliance issues during the negotiations was South Africa. Under South Africa's leadership, the delegations of Brazil, China, India, Iran, Samoa and Saudi Arabia were particularly active in shaping the views of this group on compliance.

12 The fact that non-Annex I (developing country) parties are committed to the objective, principles and process of the climate regime, but do not yet have quantified emissions-reduction commitments, is said to reflect the 'principle of common but differentiated responsibilities' as contained in Art 3.1 of the Convention, Principle 7 of the 1992 Rio Declaration on Environment and Development and many other international environmental agreements.

13 Most AOSIS members are also members of the G-77, and AOSIS worked closely with the G-77 in arriving at common positions. The AOSIS coordinator and spokesperson for compliance issues was Samoa. The author was an adviser to the delegation of Samoa on compliance.

14 Although OPEC did not negotiate as a bloc, Saudi Arabia played a leadership role in presenting the views of this group.

15 'President Bush discusses Global Climate Change', White House Press Release of 11 June 2002, available at www.whitehouse.gov.

16 The US submission envisaged a single compliance procedure divided between two institutionally-distinct standing branches or bodies, one with the responsibility for facilitation, the other for enforcement. The mandates of these bodies would be linked to the legal character of the articles of the Protocol, with the binding targets and associated commitments related to reporting and the operation of the flexibility mechanism handled by the Enforcement Branch. Soft and unquantified provisions would be handled by the Facilitative Branch. Penalties associated with non-compliance would be spelled out clearly and in advance. It was specifically proposed that the main penalty for an Annex I party exceeding its cap should be that 'any excess tonnes be subtracted from a Party's assigned amount for the subsequent commitment period, with a penalty (at a rate designed to make overages unattractive)'. September 1999 Submissions, 65–81.

17 The JWG was chaired by Espen Ronneberg (Republic of the Marshall Islands) and Harald Dovland (Norway). Mr Ronneberg was later replaced by Ambassador Tuiloma Neroni Slade (Samoa). The co-chairs were supported throughout by the staff of the UNFCCC Secretariat, Mukul Sanwal and Xueman Wang, and Monica Sevilla.

18 A final push in the development of the compliance system came from the combined efforts of Jan Pronk (the Netherlands) and Valli Moosa (South Africa), the ministers tasked with pushing through compromises on critical parts of the text. For a full account of Pronk and Moosa's roles, see Lefebere, 2002.

19 See *Report of the Ad Hoc Group on Article 13 on its Sixth Session*, FCCC/AG-13/1998/2, and *Report of the Fourth Session of the Conference of the Parties to the UNFCCC*, Decision 10/CP.4 and Annex.

20 See also Victor et al, 1998. The leading text supporting the managerialist approach is frequently cited in both of these studies as Chayes and Chayes, 1996.

21 Both the MCP and the Kyoto Protocol take into account the possibility that the MCP's jurisdiction could be extended to cover Protocol commitments.

22 See, for example, the submission by Australia, in April 1999 Submissions, 3; and by South Africa on behalf of the Group of 77 and China, April 1999 Submissions, add 3, 3.

23 Joint Implementation under Art 6, the Clean Development Mechanism under Art 12, and emissions trading under Art 17.

24 See, for example-, the submission of Canada in September 1999 submissions, 14.

25 US Clean Air Act sec 403(b), 40 CFR sec 72.2.

26 A number of US interventions and submissions stressed the importance of strong domestic enforcement systems in meeting the Kyoto targets, and called upon parties to report on these systems in detail. See September 1999 Submission, 79.

27 Article 4 allows members of the European Community (or any other group of Parties) to enter into an agreement to jointly fulfill their commitments under the Protocol by collectively sharing these commiments as long as their combined emissions do not exceed their combined assigned amounts.

28 See, for example, 'Russia's own poor health leaves bounty of "hot air" to burn for cash: Emissions monitoring still poor, but credits could yield riches', in *The World Paper* ONLINE, March 2000, www.worldpaper.com.

29 See, for example, the submission of South Africa on behalf of the Group of 77 and China, April 1999 Submissions.

30 See, eg, the submission of South Africa on behalf of the Group of 77 and China, April 1999 Submissions, which calls for consideration of the benefits behind automatic penalties, contemplates the use of financial penalties, and states clearly that 'binding consequences for non-compliance are essential'.

31 Submissions were received from Argentina, Brazil, India, Samoa (for AOSIS), Saudi Arabia and South Africa.

32 See, for example, the submissions of Brazil, AOSIS and South Africa in 2000 Submissions.

33 Submission of Brazil, 2000 Submissions 24–27. Brazil's position was supported in part by a reading of Art 8.3 of the Protocol, which envisages a role for the COP/MOP in receiving the reports of the ERTs prior to the identification by the Secretariat of the questions of implementation contained in those reports, which would then be considered further by the COP/MOP. Brazil and others read this provision as authorizing the COP/MOP to review and filter these questions before any further action (including action by the Compliance Committee) could be taken.

34 See, for example, Submission of Argentina, 2000 Submissions, 9, supporting a role for the COP/MOP in vetting the final report of the Compliance Committee; Submission of China, 43, supporting a role for the COP/MOP in triggering the Compliance Procedure.

35 This would follow the lines of the role of the WTO Dispute Settlement Body in relation to the Panels and the Appellate Body in that system. One member of the subcommittee, by blocking consensus, could ensure that the decision of the Compliance Committee would stand. Submission of Brazil, 2000 Submissions, 25.

36 For example, at COP-1, when the parties adopted the Berlin Mandate that committed them to negotiate a Protocol containing targets and timetables, Kuwait, Saudi Arabia and Venezuela formally expressed their reservations on the decision adopted. FCCC/CP/1995/7. At COP-2, Bahrain, Jordan, Kuwait, Nigeria, Oman, Qatar, the Russian

*The Negotiation of a Kyoto Compliance System* 35

19 See *Report of the Ad Hoc Group on Article 13 on its Sixth Session*, FCCC/AG-13/1998/2, and *Report of the Fourth Session of the Conference of the Parties to the UNFCCC*, Decision 10/CP.4 and Annex.

20 See also Victor et al, 1998. The leading text supporting the managerialist approach is frequently cited in both of these studies as Chayes and Chayes, 1996.

21 Both the MCP and the Kyoto Protocol take into account the possibility that the MCP's jurisdiction could be extended to cover Protocol commitments.

22 See, for example, the submission by Australia, in April 1999 Submissions, 3; and by South Africa on behalf of the Group of 77 and China, April 1999 Submissions, add 3, 3.

23 Joint Implementation under Art 6, the Clean Development Mechanism under Art 12, and emissions trading under Art 17.

24 See, for example-, the submission of Canada in September 1999 submissions, 14.

25 US Clean Air Act sec 403(b), 40 CFR sec 72.2.

26 A number of US interventions and submissions stressed the importance of strong domestic enforcement systems in meeting the Kyoto targets, and called upon parties to report on these systems in detail. See September 1999 Submission, 79.

27 Article 4 allows members of the European Community (or any other group of Parties) to enter into an agreement to jointly fulfill their commitments under the Protocol by collectively sharing these commiments as long as their combined emissions do not exceed their combined assigned amounts.

28 See, for example, 'Russia's own poor health leaves bounty of "hot air" to burn for cash: Emissions monitoring still poor, but credits could yield riches', in *The World Paper* ONLINE, March 2000, www.worldpaper.com.

29 See, for example, the submission of South Africa on behalf of the Group of 77 and China, April 1999 Submissions.

30 See, eg, the submission of South Africa on behalf of the Group of 77 and China, April 1999 Submissions, which calls for consideration of the benefits behind automatic penalties, contemplates the use of financial penalties, and states clearly that 'binding consequences for non-compliance are essential'.

31 Submissions were received from Argentina, Brazil, India, Samoa (for AOSIS), Saudi Arabia and South Africa.

32 See, for example, the submissions of Brazil, AOSIS and South Africa in 2000 Submissions.

33 Submission of Brazil, 2000 Submissions 24–27. Brazil's position was supported in part by a reading of Art 8.3 of the Protocol, which envisages a role for the COP/MOP in receiving the reports of the ERTs prior to the identification by the Secretariat of the questions of implementation contained in those reports, which would then be considered further by the COP/MOP. Brazil and others read this provision as authorizing the COP/MOP to review and filter these questions before any further action (including action by the Compliance Committee) could be taken.

34 See, for example, Submission of Argentina, 2000 Submissions, 9, supporting a role for the COP/MOP in vetting the final report of the Compliance Committee; Submission of China, 43, supporting a role for the COP/MOP in triggering the Compliance Procedure.

35 This would follow the lines of the role of the WTO Dispute Settlement Body in relation to the Panels and the Appellate Body in that system. One member of the subcommittee, by blocking consensus, could ensure that the decision of the Compliance Committee would stand. Submission of Brazil, 2000 Submissions, 25.

36 For example, at COP-1, when the parties adopted the Berlin Mandate that committed them to negotiate a Protocol containing targets and timetables, Kuwait, Saudi Arabia and Venezuela formally expressed their reservations on the decision adopted. FCCC/CP/1995/7. At COP-2, Bahrain, Jordan, Kuwait, Nigeria, Oman, Qatar, the Russian

*The Negotiation of a Kyoto Compliance System* 35

19 See *Report of the Ad Hoc Group on Article 13 on its Sixth Session*, FCCC/AG-13/1998/2, and *Report of the Fourth Session of the Conference of the Parties to the UNFCCC*, Decision 10/CP.4 and Annex.

20 See also Victor et al, 1998. The leading text supporting the managerialist approach is frequently cited in both of these studies as Chayes and Chayes, 1996.

21 Both the MCP and the Kyoto Protocol take into account the possibility that the MCP's jurisdiction could be extended to cover Protocol commitments.

22 See, for example, the submission by Australia, in April 1999 Submissions, 3; and by South Africa on behalf of the Group of 77 and China, April 1999 Submissions, add 3, 3.

23 Joint Implementation under Art 6, the Clean Development Mechanism under Art 12, and emissions trading under Art 17.

24 See, for example-, the submission of Canada in September 1999 submissions, 14.

25 US Clean Air Act sec 403(b), 40 CFR sec 72.2.

26 A number of US interventions and submissions stressed the importance of strong domestic enforcement systems in meeting the Kyoto targets, and called upon parties to report on these systems in detail. See September 1999 Submission, 79.

27 Article 4 allows members of the European Community (or any other group of Parties) to enter into an agreement to jointly fulfill their commitments under the Protocol by collectively sharing these commiments as long as their combined emissions do not exceed their combined assigned amounts.

28 See, for example, 'Russia's own poor health leaves bounty of "hot air" to burn for cash: Emissions monitoring still poor, but credits could yield riches', in *The World Paper* ONLINE, March 2000, www.worldpaper.com.

29 See, for example, the submission of South Africa on behalf of the Group of 77 and China, April 1999 Submissions.

30 See, eg, the submission of South Africa on behalf of the Group of 77 and China, April 1999 Submissions, which calls for consideration of the benefits behind automatic penalties, contemplates the use of financial penalties, and states clearly that 'binding consequences for non-compliance are essential'.

31 Submissions were received from Argentina, Brazil, India, Samoa (for AOSIS), Saudi Arabia and South Africa.

32 See, for example, the submissions of Brazil, AOSIS and South Africa in 2000 Submissions.

33 Submission of Brazil, 2000 Submissions 24–27. Brazil's position was supported in part by a reading of Art 8.3 of the Protocol, which envisages a role for the COP/MOP in receiving the reports of the ERTs prior to the identification by the Secretariat of the questions of implementation contained in those reports, which would then be considered further by the COP/MOP. Brazil and others read this provision as authorizing the COP/MOP to review and filter these questions before any further action (including action by the Compliance Committee) could be taken.

34 See, for example, Submission of Argentina, 2000 Submissions, 9, supporting a role for the COP/MOP in vetting the final report of the Compliance Committee; Submission of China, 43, supporting a role for the COP/MOP in triggering the Compliance Procedure.

35 This would follow the lines of the role of the WTO Dispute Settlement Body in relation to the Panels and the Appellate Body in that system. One member of the subcommittee, by blocking consensus, could ensure that the decision of the Compliance Committee would stand. Submission of Brazil, 2000 Submissions, 25.

36 For example, at COP-1, when the parties adopted the Berlin Mandate that committed them to negotiate a Protocol containing targets and timetables, Kuwait, Saudi Arabia and Venezuela formally expressed their reservations on the decision adopted. FCCC/CP/1995/7. At COP-2, Bahrain, Jordan, Kuwait, Nigeria, Oman, Qatar, the Russian

Federation, Saudi Arabia, Sudan, the Syrian Arab Republic, the United Arab Emirates, Venezuela and Yemen, and one observer state, the Islamic Republic of Iran, formally objected to the adoption, approval or acceptance of the Geneva Ministerial Declaration which endorsed the latest assessment report of the IPCC and reiterated the call for swift conclusion of the Kyoto Protocol negotiations. FCCC/CP/1996/15, Annex IV.

37 Art 2.3, 3.14.

38 By COP-7, the Umbrella group had basically fallen apart, with those countries generally supportive of a strong compliance system with binding consequences (Norway, Iceland and New Zealand, known as 'Nizeland') agreeing largely with the G-77, EU+ and the Environmental Integrity Group (Mexico, South Korea and Switzerland). What remained of the Umbrella group (Australia, Japan, Canada, Russia and the US, referred to here as the Umbrella rump) had either rejected the concept of binding consequences, rejected the Protocol altogether, or were showing solidarity with allies.

39 The US never publicly argued against the need for binding consequences.

40 While the idea of financial consequences had appeared in a number of G-77 and EU submissions, it became part of the negotiating package as a result of a proposal from Valli Moosa (South Africa). Under the two main proposals on the table, one from the EU and one from Brazil, the fines collected would be placed in a Compliance Fund, which would invest in generating emissions reductions. Fund proponents had not yet discussed fully or resolved positions on whether such a Compliance Fund should be managed by an institution (including a domestic institution) chosen by the party in non-compliance (EU), or by an international institution designated by the COP/MOP or by the Compliance Committee (Brazil). The Brazilian and EU proposals left unresolved whether the Fund's investments should be exclusively domestic within the non-complying party, whether they should be global (Brazil) or instead should be left to the discretion of the Fund (EU). Minister Moosa introduced language that would have obligated parties in non-compliance to provide 'reparation of damage to the environment'.

41 See Chapter 3 for a discussion of how Annex I interests were protected in the context of this majority membership, using other procedural means such as presumptions and voting rules.

42 Marrakesh Accords, Decision 15/CP.7, Annex, FCCC/CP/2001/13/Add.2, para 5.

43 Text proposed by the co-chairmen of the JWG on Compliance, FCCC/SB/2000/11, 24 October 2000.

44 Marrakesh Accords, Decision 24/CP.7, FCCC/CP/2001/13/Add.3.

## References

Bodansky, D. (2001) 'Bonn Voyage: Kyoto's Uncertain Revival', *The National Interest*, vol 45, fall

Chayes, A. and Chayes, A. H. (1996) *The New Sovereignty: Compliance with International Regulatory Agreements*, Harvard University Press, Cambridge, MA

Danish, K. (1997) 'Management v Enforcement: The New Debate on Promoting Treaty Compliance', *Virginia Journal of International Law*, vol 37, spring, p789

Downs, G., Danish, K. and Barsoom, P. (2000) 'The Transformational Model of International Regime Design: Triumph of Hope or Experience?', *Columbia Journal of Transnational Law*, vol 38, p465

Koskenniemi, M. (1992) 'Breach of Treaty or Non-Compliance? Reflections on the Enforcement of the Montreal Protocol', *Yearbook of International Environmental Law*, vol 3, p123

Lefebere, R. (2002) 'From The Hague to Bonn to Marrakesh and Beyond: A Negotiating History of the Compliance System under the Kyoto Protocol' in Kiss, A. C. (ed) *Hague Yearbook of International Law 2001*, Kluwer Law International, The Hague, London and New York, pp25–54

Szell, P. (1995) 'The Development of Multilateral Mechanisms for Monitoring Compliance' in Lang, W. (ed) *Sustainable Development and International Law*, Kluwer Law International, The Hague, London and New York

Victor, D. (1998) 'The Operation and Effectiveness of the Montreal Protocol Implementation Committee' in Victor, D. G. et al (eds) *The Implementation and Effectiveness of International Environmental Commitments: Theory and Practice*, MIT Press, Cambridge, MA, p137

Victor, D. G., Raustiala, F. and Skolnikoff, E. B. (eds) (1998) *The Implementation and Effectiveness of International Environmental Commitments: Theory and Practice*, MIT Press, Cambridge, MA, pp681–2

Werksman, J. (1996) 'Compliance and Transition: Russia's Non-Compliance Tests the Ozone Regime', *Zeitschrift für ausländisches öffentliches Recht und Völkerrecht*, vol 56, p750

Werksman, J. (1999) 'Compliance and the Kyoto Protocol: Building a Backbone into a Flexible Regime' in Brunnee, J. and Hay, E. (eds) *Yearbook of International Environmental Law*, vol 9, Oxford University Press, Oxford, pp48–101

Chapter 2

# The Kyoto Compliance System: Towards Hard Enforcement

Geir Ulfstein and Jacob Werksman

## Introduction

As states take on more comprehensive and stricter obligations in multilateral environmental agreements (MEAs), increasing focus is being given to the effective implementation of these obligations. When the world's environment ministers met in Malmö in 2000 they stated that there is an 'alarming discrepancy between commitments and action' and referred to 'the central importance of environmental compliance, enforcement and liability'[1] in correcting that discrepancy. The United Nations Environment Programme (UNEP)'s Governing Council called in 2001 for 'speedy implementation of the legal commitments contained in the multilateral environmental agreements' and requested UNEP's Executive Director to continue the work with the 'draft guidelines on compliance with multilateral environmental agreements'.[2] Following up on this mandate, a special session of UNEP's Governing Council adopted a set of guidelines on compliance with and enforcement of MEAs.[3] These guidelines recognize, however, that the parties to each treaty have the primary responsibility for designing effective mechanisms and procedures to ensure implementation.[4]

Many MEAs call for parties to report on their implementation and a few have established specialized bodies and procedures tailor-made to deal with any cases of non-compliance that may arise (see also Chapter 10 by Wettestad).[5] A number of advantages of using such internal non-compliance mechanisms, rather than more traditional dispute settlement procedures, have been highlighted in the literature.[6] First, questions of compliance with MEA commitments are multilateral in character and may affect all parties, rather than any particular party specifically. Non-traditional non-compliance mechanisms allow compliance issues to be addressed in a multilateral context, rather than through bilateral disputes resolved through third party arbitration or adjudication. Second, non-compliance procedures can be designed to head off potential non-compliance, rather than waiting for a formal case of breach to be established. Finally, non-compliance procedures may promote the resolution of compliance problems in a cooperative, rather than adversarial, manner through procedures designed to facilitate rather than enforce compliance. It has, however, been questioned whether such non-confrontational, managerial approaches can be equally effective for those MEAs that impose environmental obligations with heavier economic and social costs.[7]

In the Kyoto Protocol context, the Marrakesh Accords have established a Compliance Committee with both a Facilitative Branch and an Enforcement Branch to address compliance with the substantive commitments. We will address both branches of the Compliance Committee. The emphasis will, however, be on the 'hard' elements of Kyoto's compliance system, i.e. the sanctions or enforcement consequences applied in cases of breaches of obligations, and the procedures that will determine and apply such consequences.

As has been described in Chapter 1, the agreement by negotiators to strengthen the consequences that would attach to non-compliance was reached through the inclusion of institutional and procedural elements designed to promote due process. The concept of due process may be considered to reflect elements of both effectiveness and fairness. Due process describes procedures and institutions that are appropriate to the task at hand and that ensure, for example, that the decision-making body has access to the information, expertise and authority necessary to take the decisions assigned to it. More often, the concept of due process is used to describe the treatment that interested parties, and in particular the party concerned, are entitled to receive as a matter of fairness and justice. Fairness is both an absolute and a relative standard, as the party concerned will assess its own treatment in comparison to the way in which other parties in comparable positions have been treated.

These two aspects of due process – effectiveness and fairness – can come into tension in any compliance system. However, in the international context, where sovereigns remain strong and techniques for enforcement are weak, it is all the more important that parties found to be in non-compliance perceive that they have been treated fairly and objectively, so that arguments about the illegitimacy of the institutions and procedures are removed as an excuse for ignoring decisions taken against them. Perceptions of fairness based, for example, on the geographical balance of the membership of a decision-making body can be as important in this regard as the actual even-handedness of the decision makers.[8]

This chapter provides a detailed analysis of the institutions, procedures and consequences designed for the Kyoto compliance system, with a particular focus on whether the negotiators succeeded in building an enforcement regime that combines fairness and effectiveness in a manner that is likely to deliver due process. Given the uniqueness of the climate change regime, this assessment is specific to that context. Precedents within other international and domestic systems, however, inevitably influence our perceptions of appropriate standards of due process. At the international level, our main points of comparison are those regimes that most directly influenced the Kyoto negotiators: the Montreal Protocol's non-compliance procedure, the World Trade Organization (WTO)'s dispute settlement system, and the rules and procedures of the International Court of Justice (ICJ). Domestic analogues to the Kyoto system are more difficult to construct. The Kyoto system is non-adversarial, and focuses on technical determinations of compliance rather than criminal guilt or civil liability. In this sense it resembles an administrative, rather than a criminal or civil procedure. On the other hand, the compliance system's enforcement measures and its due process guarantees are comparable to those applied by judicial bodies.

As with most administrative procedures, due process under Kyoto's Enforcement Branch is focused primarily on ensuring that decisions are based on accurate and unbiased information, follow rational procedures, and are shielded from inappropriate political influence. The procedural guarantees necessary to achieve these goals will depend in large part on the nature of the specific decisions that are being taken and, in particular, the degree of discretion left to the decision makers with regard to the facts and the law before them. As was discussed in Chapter 1, this issue was characterized by the Kyoto system's negotiators as striking a balance between automaticity and discretion.

Some delegations argued that due process would be better served through the certainty of an automatic process, whereby predetermined consequences would be triggered by pre-defined categories of breaches. Such an automatic process requires that precisely drafted rules be applied to uncontested information about a party's performance. This places the burden on negotiators to design a system that all parties perceive as fair, in which both absolute and relative fairness are achieved because all parties in comparable situations are treated identically. In this model, the goals of due process are achieved through the *absence* of process, which is rendered unnecessary by an automatic application of the rules.

Others argued that fairness could only be achieved through a case-by-case examination of the 'type, degree and causes' of non-compliance of the party concerned, as is suggested in the language of Article 18 of the Protocol. This view was informed, in part, by an assessment that it was unlikely that negotiators had the prescience to design rules with sufficient precision to obviate the need for case-by-case judgements. Fairness would require the exercise of a degree of flexibility and discretion, and the application of experience learned in the operation of the treaty.

The enforcement process ultimately agreed in Marrakesh combines elements of automaticity and discretion, depending on the nature of the decision being taken, and the institution and procedure entrusted with taking that decision.

The next section provides a general overview of the instititutional set-up of the compliance system established by the Marrakesh Accords, including the role of the Expert Review Teams (ERTs), the Compliance Committee and its Facilitative Branch. We then describe in more detail the composition, mandate and procedures that will govern the operation of the Enforcement Branch. Finally, we turn to the nature of the enforcement-related decisions entrusted to the Enforcement Branch, to assess the manner in which the text has balanced automaticity and discretion, fairness and effectiveness, in providing due process to the climate change regime.

## The institutional framework

As is described in other chapters in this volume, there are two main institutional arrangements involved in the identification of non-compliance by Annex I parties: the Compliance Committee and the ERTs, which are tasked with conducting a 'thorough and comprehensive technical assessment' of the performance of each Annex I party to the Protocol.

The size, composition, capacity and competence of these bodies are essential determinants of due process. Size matters, because the larger the body the more representative and knowledgeable it can be of the circumstances of a wide range of parties. However, if a body is too large, it can become unwieldy and bureaucratic, or it may become overly reliant on political processes to reach decisions. Smaller bodies, on the other hand, may unduly concentrate power in the hands of a limited number of individuals.

As has been described in Chapter 1, the balance in composition of the Enforcement Branch was a particularly controversial issue in the negotiations.[9] In many UN institutions, balance in composition is assessed in terms of geographical balance among the five traditional UN groupings (Africa, Asia, Eastern Europe, Latin America and the Caribbean, and the Western European and Others Group (WEOG)). In the context of the climate change, balance is also viewed in terms of the balance between Annex I and non-Annex I parties, as well as the inclusion of parties with special interests in climate change: the particularly vulnerable small island developing countries, and countries with economies particularly dependent on the export of fossil fuels. Some delegations raised due process concerns with respect to Enforcement Branch composition, suggesting that individuals from non-Annex I parties, which do not have quantified commitments under the Protocol, should not sit in judgement of Annex I party performance. These arguments were sometimes justified on the basis of analogies to the common law jury system, which, in some circumstances, entitles defendants to a jury of their peers. From a legal perspective, the analogy is weak for a number of reasons. Members of the ERTs and the Compliance Committee are acting as experts making technical judgements, not as citizens asked to apply common sense to test the veracity of facts and witnesses. Furthermore, even jury systems do not select peers by including only members that are under the same legal obligations as the accused; a doctor accused of medical negligence is not guaranteed a jury of fellow physicians. Indeed, loading juries in such a way might have the effect of biasing judgments towards the interests of the accused, and neglecting the interests of the victims. The analogy, however weak in a legal sense, is nonetheless relevant, as it does reflect some delegations' political perceptions of unfairness, which, as has been indicated, could affect the acceptance and enforceability of decisions.

'Capacity' refers to the capacity in which members of the body are expected to serve: either personal and professional, or as the representatives of governments or groupings of governments, acting upon political instructions. Ensuring that members act in their personal capacities can help to make more palatable the political flavour given to a body through an insistence on geographical balance. If members act in a personal capacity, it is important that they have a clear understanding that although they may be filling a particular geographical quota, they have been appointed to provide a particular perspective, rather than to advocate on behalf of a particular geographical grouping.

The competence of a body will depend upon the professional qualifications demanded of its members. Such qualifications must be tailored to the task at hand, but be sufficiently flexible so as to ensure that professional, cultural and educational biases do not act to undermine a body's representativeness. As will be seen, given the complexity of the climate change issue, the range of expertise that could be seen as relevant to assessing compliance with the Protocol is wide, spanning the natural sciences, economics, political science and law.

Finally, the means by which the criteria of composition, capacity and competence are applied is important. If the selection of candidates for these posts is turned over to a purely political process, agreed criteria could simply be ignored. One issue of concern is the tendency for the individual delegates who have designed these Protocol institutions to seek appointment to them as members. For example, many of the members of the Executive Board of the Kyoto Protocol's Clean Development Mechanism (CDM), who are expected to act on the board in their personal capacity, also participated in its design, and continue to represent their governments in the ongoing negotiations on the design of the CDM and in other aspects of the Protocol. While these individuals are undoubtedly experts in the texts they themselves designed, it is open to question whether it is reasonable to expect them to wear both hats, of independent decision maker and diplomatic representative, comfortably.

## *Expert Review Teams*

ERTs will play an innovative and important part in the enforcement of the climate commitments, building on the experience of the use of In-Depth Review teams under the UNFCCC. Although ERTs do not have a mandate to make determinations of non-compliance, they are called upon to identify 'questions of implementation' with regard to a party's performance, by providing a 'technical assessment' and 'identifying any potential problems in, and factors influencing, the fulfilment of commitments'.[10] As will be discussed, the Kyoto compliance system also allows questions of implementation to be raised by any party with regard to any other party, and by a party with regard to itself. It is nonetheless widely anticipated that most questions presented to the Compliance Committee will be contained in the reports of the ERTs. Ensuring that the ERTs operate in an effective and unbiased manner will therefore be essential to the system's integrity.

ERTs are to be 'coordinated by the secretariat' and 'composed of experts selected from those nominated by Parties to the Convention and, as appropriate, by intergovernmental organizations'.[11] The Marrakesh Accords set out additional details on the composition, qualifications and work of the ERTs.[12] The credibility of the ERTs lies first of all in their composition as a team of independent experts. A neutral organ, the UNFCCC Secretariat, composes each team, which is selected from a list of experts nominated by the parties. While such nominations may not necessarily secure the best and the most independently minded persons, giving the parties a say in the composition of the ERTs may also contribute to the parties' trust in the process.

Experts are required to serve in their personal capacity[13] and 'refrain from making any political judgement'.[14] Furthermore, to promote the political and financial independence of any ERT, the experts selected to review a particular country's implementation 'shall neither be nationals of the Party under review, nor be nominated or funded by that Party'.[15] The requirement that there shall be a 'balance between experts from Annex I and non-Annex I Parties in the overall composition of the Expert Review Teams'[16] was intended to secure the involvement and support of non-Annex I parties in the review process. This effort at balance recognizes that there is relevant expertise in many developing countries, and that the involvement of non-Annex I experts can help further develop such expertise. These arguments helped overcome objections based on the jury analogy that would have dropped any requirement to include experts from parties that had not undertaken specific commitments.

With regard to the ERT's competence, experts shall have 'recognized competence in the areas to be reviewed' under the relevant guidelines. Training, and subsequent assessment of qualifications, shall be provided in accordance with decisions from the Conference of the Parties (COP) and the Conference of the Parties serving as the Meeting of the Parties (COP/MOP).[17] It might be asked, to what extent do the ERTs have competence to point out any potential problems without the inclusion of experts on international law? The parties may have decided not to expressly request competence in this field because conclusions from such experts might prove embarrassing for parties under review, and prevent their cooperation. This decision could also be defended on the basis that ERTs are only to identify potential problems, and not determine whether Protocol commitments are violated.[18]

The basis for the ERTs' assessment of the parties' implementation will be reports from the parties and in-country visits. In addition, ERTs 'may put questions to, or request additional or clarifying information' from the parties regarding any potential problem identified by the team.[19] Parties 'should' provide the ERTs with 'access to information necessary to substantiate and clarify the implementation of their commitments', and 'make every reasonable effort to respond to all questions and requests from the Expert Review Teams for additional clarifying information'.[20] Although 'should' is usually taken only to refer to a non-binding commitment, a political commitment is at least indicated through this agreed language. It would also seem that since the ERTs are under an obligation to provide a 'thorough and comprehensive' technical assessment, they must use all information available, including information from sources other than the relevant party. The ERTs shall, in their report, indicate the sources of information used.[21]

### *The Compliance Committe*

The procedures and mechanisms relating to compliance under the Kyoto Protocol were adopted as Decision 24/CP.7 of the Marrakesh Accords, and were recommended for adoption at the first meeting of the COP/MOP. A Compliance Committee of 20 members will be established, to function through a Plenary, a Bureau and two branches: a Facilitative Branch and an Enforcement Branch.[22] The Committee meeting as a whole will not address individual cases of non-compliance, but will report on the Committee's activities and submit proposals on administrative and budgetary matters to the COP/MOP.[23] The COP/MOP may consider such reports, provide general policy guidance, etc., but may not engage in individual non-compliance cases, except in its limited competence to address cases appealed (see below).[24] The Secretariat has also a limited role, in acting as a link between the parties and the ERTs on the one hand, and with the Compliance Committee on the other.

This leaves the handling of decisions relating to individual cases to the Bureau and the two branches of the Committee. The Bureau's role is to allocate questions of implementation to either the Facilitative Branch or the Enforcement Branch.[25] While the decision to allocate may appear to be significant, the discretion of the Bureau is, in fact, limited by the respective mandates of each Branch. When read together, these mandates make clear that the Enforcement Branch has exclusive jurisdiction over questions of implementation relating to targets that arise after the end of the first commitment period, and, with respect to methodological, reporting and eligibility requirements for participation in the flexibility mechanisms, after the beginning of the first commitment period. All other questions of implementation fall within the jurisdiction of the Facilitative Branch. Thus the timing and the substantive content of the commitment should render the process of allocation automatic.

## *The Facilitative Branch*

Ten of the 20 members of the Compliance Committee will comprise the Facilitative Branch of the Committee, elected by the COP/MOP.[26] The Facilitative Branch consists of one member from each of the five regional groups of the United Nations, one member from the Small Island Developing States (SIDS), two members from Annex I parties, and two members from non-Annex I parties.[27] All members serve in their individual capacities.[28] The arrangements for the Facilitative Branch's election, composition and capacity are similar to those for the Enforcement Branch, which are further discussed below.

The expertise required by the members of the Facilitative Branch is 'recognized competence relating to climate change and in relevant fields such as the scientific, technical, socio-economic or legal fields'.[29] These different competences shall be sought to be reflected in a 'balanced manner' within the Facilitative Branch.[30] Unlike the Enforcement Branch, there is no mandatory requirement of legal expertise among all members, as the task of this branch is facilitation rather than enforcement.

The Facilitative Branch is responsible for providing 'advice and facilitation to parties in implementing the Protocol, and for promoting compliance with Protocol commitments', taking into account 'the principle of common but differentiated responsibilities and respective capabilities' and relevant 'circumstances pertaining to the question before it'. [31] The Facilitative Branch does not make legally binding determinations of non-compliance.[32]

The Facilitative Branch is also expressly responsible for questions of implementation, falling outside the mandate of the Enforcement Branch, which relate to the minimization of adverse effects on developing countries by developed countries' implementation of their Protocol commitments,[33] and that relate to information provided on the use by Annex I parties of the flexibility mechanisms as supplementary to domestic action.[34]

Finally, in order to promote compliance and provide for 'early warning of potential non-compliance', the Facilitative Branch is tasked to address questions concerning the quantified emissions limitation and reduction commitments of Annex I parties prior to and during the commitment period,[35] commitments relating to the establishment of national systems for the estimation of emissions of greenhouse gases by sources and removal by sinks prior to the first commitment period, and commitments on the reporting of supplementary information prior to the beginning of the first commitment period.[36] These responsibilities are designed not to overlap with the responsibilities of the Enforcement Branch. It is of particular importance that the Facilitative Branch will not deal with non-compliance concerning the parties' implementation of emission limitation or reduction commitments after the end of the commitment period.

In reaching decisions, the procedural requirements of the Facilitative Branch are the same as those used by the Compliance Committee, i.e. to try to reach consensus, but if that proves impossible, to adopt decisions by a majority of at least three-quarters of the members present and voting.[37] Unlike the Enforcement Branch, there is no requirement of a double majority among Annex I and non-Annex I parties.

The Facilitative Branch may decide upon the application of one or more of a prescribed list of consequences, taking into account 'the principle of common but differentiated responsibilities and respective capabilities'. These consequences consist of advice and facilitated assistance, facilitation of financial and technical assistance, and the formulation of recommendation.[38]

## The Enforcement Branch

### *Composition and mandate*

Like the Facilitative Branch, the Enforcement Branch consists of ten members of the 20-member Compliance Committee.[39] This size reflects a negotiated compromise between concerns for having an effective smaller body on the one hand, and a desire for a bigger, more representative but perhaps less effective body on the other. The limited number of members means in itself that not every party will be represented, which serves to increase the body's independence in practice.

Only once before has the UNFCCC established a standing body of limited membership – the Bureau of the Conference of the Parties to the Convention. The Bureau proved, in the end, to be one of the models on which the Enforcement Branch (and the Compliance Committee as a whole) was based. Under the rules of procedure for the COP, each of the five regional groups is represented on the Bureau by two members, and one Bureau member represents the SIDS.[40] The Bureau's composition, based on geographical spread, is not without precedent in dispute settlement bodies. The United Nations regional groups are also used, for example, in electing judges to the ICJ, though the permanent members of the Security Council in practice each will have a judge of their nationality. The Montreal Protocol Implementation Committee consists of ten parties 'based on equitable geographical distribution'.[41] Similarly, the WTO Appellate Body membership is to be broadly representative of membership in the WTO.[42]

In the end, in creating the Enforcement Branch, delegations agreed on a single formula for composition that applied to all of the bodies of limited size created by the Protocol as follows:

1 One member from each of the five regional groups of the United Nations and one member from the SIDS, taking into account the interest groups as reflected by the current practice in the Bureau of the COP;

2 two members from parties included in Annex I; and
3 two members from parties not included in Annex I.

The reference to the 'current practice in the Bureau' indicates that an effort should be made to ensure that at any time, one of the seats held by a developing country regional group should be occupied by a developing country whose economy is highly dependent on income generated from the production, processing and export of fossil fuels.

The formula yields the following division between Annex I parties, which are subject to the Enforcement Branch's jurisdiction, and non-Annex I parties, which are not:

**Table 2.1 Enforcement branch membership**

| | **Annex I** | **Non-Annex I** |
|---|---|---|
| Annex I | 2 | |
| Non-Annex I | | 2 |
| African Group | | 1 |
| Asia Pacific Group[43] | | 1 |
| Eastern European Group | 1 | |
| Group of Latin America and the Caribbean (GRULAC) | | 1 |
| Small Island Developing States (SIDS) | | 1 |
| Western European and Others Group (WEOG) | 1 | |
| **Totals** | **4** | **6** |

Enforcement Branch members, like all other members of the Compliance Committee, are to 'serve in their individual capacities'.[44] This formal independence from instruction by their country of nationality – or any other bond – is intended to strengthen the Enforcement Branch's credibility. However, unlike the ERTs, which are selected by the apolitical secretariat, the members of the Enforcement Branch are elected by the COP/MOP. This opens the possibility for taking non-professional factors, especially political factors, into account. Elections by political organs are known both in national courts and in international tribunals; for example, the ICJ, the Montreal Protocol and the WTO Appellate Body all elect judges through the political organs of the international body.

While the members of the Compliance Committee shall have 'recognized competence relating to climate change and in relevant fields such as the scientific, technical, socio-economic or legal fields',[45] the members of the Enforcement Branch are additionally to have 'legal experience'.[46] Such legal expertise is consistent with the function of the Enforcement Branch in 'determining' cases of non-compliance.[47] However, the level of this legal expertise was left intentionally broad to permit the COP to allow non-lawyers to serve as members. Comparison can be made to the ICJ (where the judges shall possess 'the qualifications required in their respective countries for appointment to the highest judicial offices, or are jurisconsults of recognized competence in international law'), and to the WTO Appellate Body (which requires 'persons of recognized authority, with demonstrated expertise in law, international trade and the subject matter of the covered agreements generally').[48] In contrast, panels established under the WTO Dispute Settlement Understanding may consist of non-lawyers, and the decision establishing the Montreal Protocol Implementation Committee does not require any particular expertise of its members, who are referred to as 'parties', suggesting that they serve in a diplomatic and political capacity rather than in an individual capacity.[49]

As already discussed, the mandate of the Enforcement Branch includes questions of implementation of the emissions targets by Annex I parties after the end of the first commitment period, and methodological, reporting and eligibility requirements regarding the flexibility mechanisms after the beginning of the first commitment period. This mandate is designed to cover the most central elements of the Kyoto Protocol, and avoid overlap with the mandate of the Facilitative Branch.

**Box 2.1 Mandate of the Enforcement Branch**

1 **The Enforcement Branch shall determine whether a party is in non-compliance with:**

- Its quantified emission limitation or reduction commitment under Article 3, paragraph 1, of the Protocol, at the end of the first commitment period.
- The methodological and reporting requirements under Article 5, paragraphs 1 and 2, and Article 7, paragraphs 1 and 4, of the Protocol, after the beginning of the first commitment period.
- The eligibility requirements for participation in the flexibility mechanisms under Articles 6, 12 and 17 of the Protocol, after the beginning of the first commitment period.

2 **The Enforcement Branch shall also resolve disagreements between an ERT and the party concerned with respect to:**

- Adjustments to that party's national inventory, under Article 5, paragraph 2, of the Protocol that have been proposed by the ERT.
- Corrections to a party's compilation and accounting database concerning the validity of a transaction under the Protocol's flexibility mechanisms, which have been proposed by the ERT.

Under the Kyoto compliance system, the information that may be used by the two branches is formally limited. The branches 'shall' base their deliberations on any relevant information from reports of ERTs, the party concerned, a party which has submitted a question of implementation with respect to another party, and reports from the COP, the COP/MOP and the subsidiary bodies of the Convention and the Protocol.[50] 'Competent' intergovernmental organizations (IGOs) and non-governmental organizations (NGOs) 'may' also submit relevant factual and technical information.[51] This means that the branch is required to rely upon information from 'official' sources, but reliable information submitted by competent IGOs or NGOs will also be available for consideration. Although the branches are unlikely to actively seek information in the same manner as the ERTs, they are authorized to seek expert advice.[52] This will serve to ensure that the branches also have access to competence on matters of fact.

## *Procedures*

The compliance system's procedures are designed to promote effectiveness in decision making and afford due process protections to concerned parties. As already indicated, due process guarantees have been strengthened when hard enforcement consequences are involved. The Marrakesh Accord's compliance procedures, set out in Sections VII, VIII and IX of Decision 24/CP.7, seek to guarantee due process by:

1 setting time limits for decisions;
2 requiring a preliminary examination of the question of implementation;
3 requiring preliminary findings to be communicated to the party concerned for comment;
4 requiring notification to the party at the different stages of the process;
5 making information available to the party;
6 allowing the party to designate persons to represent it;
7 allowing comments from the party;
8 allowing the concerned party to request a hearing; and
9 requiring decisions to include conclusions and reasons.

More restrictive time limits are set in Section X on expedited procedures in cases of suspension of eligibility to use the flexibility mechanisms in Articles 6, 12 and 17. These stricter time limits are designed to protect parties' interests in returning to the market as quickly as possible.

A significant difference between the compliance system under the Kyoto Protocol and that employed by several international courts is that there is no particular arrangement within the Protocol's process to address situations in which a member of the Enforcement Branch is of the nationality of a party involved in a compliance proceeding, or where there are other *prima facie* reasons for questioning a member's objectivity. The ICJ provides, for example, for the appointment of an *ad hoc* judge to allow both parties to a dispute to be represented on the bench. In contrast, in disputes before the WTO, citizens of Member States whose governments are parties to a dispute are excluded from WTO Panels unless the parties to the dispute agree otherwise.[53] The failure of the compliance system to deal with this issue might be seen as a weakness of the Enforcement Branch. On the other hand, it may reflect the implicit trust that the designers have placed in the independent status of the members of the two branches, or the fact that the Enforcement Branch was not considered to be the equivalent of a court or an arbitral panel.

The voting regulations present a special feature of the Compliance Committee's procedure, and particularly the Enforcement Branch's regulations. The Committee is required to make all efforts to reach consensus, but if this fails, decisions may be made by a three-quarters majority of the members present and voting.[54] The fact that substantive decisions may be taken by majority voting rather than by consensus contributes to effective decision making, and is itself remarkable, considering the continuing impasse over majority voting on substantive issues in the Rules of the Procedure in the climate change regime. However, a decision by the Enforcement Branch requires also a double majority, i.e., a majority among both Annex I and non-Annex I parties. This system seeks to accommodate the concern that Annex I parties should not be judged by the non-Annex I parties' majority in the Enforcement Branch. It may, however, be argued that the double majority requirement will not provide a fully satisfactory guarantee for Annex I parties: while they may be able to block a decision of non-compliance, the shaming effect of a three-quarters majority remains.

It could be asked whether a party accused of non-compliance should benefit from the principle afforded to individuals under penal procedure: that guilt must be proven beyond reasonable doubt. Although no such requirement is explicitly articulated in the Marrakesh procedures, nevertheless, much the same burden of proof has been structured to provide protections for a party accused of non-compliance, through provisions on self-reporting, review by ERTs and the Compliance Committee's due process guarantees, though the double majority voting requirement stops short of requiring complete consensus. Regardless, the decisions of the Enforcement Branch are not formal decisions of guilt.

In addition, there is a limited possibility for appeal to the COP/MOP against final decisions of the Enforcement Branch on due process grounds, where these decisions relate to Article 3 (1) of the Kyoto Protocol, i.e. the provision on quantified limitation and reduction of greenhouse gas emissions. Only aspects relating to denial of due process may be appealed, which suggests that the assessment of the factual evidence, the legal interpretation applied to that evidence or the consequences applied, are not all subject to challenge.[55] Finally, a three-quarters majority in the COP/MOP is required to override a decision of the Enforcement Branch, and the COP/MOP may only refer the case back to the Enforcement Branch. It may not make its own decision on whether non-compliance has occurred, and what may be the relevant consequences.[56]

Although it might be argued that a broader scope for appeal would provide additional due process guarantees, this is only the case if such appeals are addressed to a competent judicial body, rather than to a political body such as the COP/MOP. The reduction of the political role of the COP/MOP is thus itself a guarantee of due process. The limitation on the scope for appeal and the requirement of the super-majority will in practice allow the COP/MOP a very limited role in individual cases of non-compliance. The requirement of a large qualified majority in order to overturn decisions of the Enforcement Branch has similarities to the WTO process, where decisions of a Panel (if they are not appealed) and of the Appellate Body will stand, unless overturned by a consensus of the WTO membership.

### Determining non-compliance

The Enforcement Branch has two main sets of tasks: determining whether a party is in non-compliance with certain obligations, and resolving disagreements between an

ERT and a party over inconsistencies in its greenhouse gas inventories, or in its system for accounting for transactions in the use of the flexibility mechanisms.

We will analyse the most serious of these tasks from a due process perspective: the assessment of whether a party has met its target by remaining within its assigned amount. This assessment by the Enforcement Branch of whether a target has been achieved can, by definition, only take place after the end of the commitment period. The first commitment period runs from 2008 to 2012, but the parties are allowed a certain grace period after the end of the commitment period, where they are allowed to acquire and transfer emission quotas in order to fulfill their targets.[57] However, the reporting requirements that will provide the basis for assessing compliance with a party's commitments will be assessed regularly throughout the commitment period. Incentives to comply with these reporting requirements are provided through the links between the reporting requirements and the eligibility of a party to participate in the Protocol's flexibility mechanisms. This should mean that by 2012–14, when compliance with targets is first assessed, the Enforcement Branch will have already acquired a good deal of experience in assessing and incentivizing a party's performance.

At first glance, the task of assessing compliance with a party's target appears to be straightforward. Each Annex I party will have had established, before the start of the commitment period, an Assigned Amount, expressed in terms of tonnes of carbon dioxide equivalent.[58] Emissions will be recorded and submitted by the party to an ERT for review on an annual basis. At the end of the commitment period, and following the additional grace period, a party's total emissions will be compared against its Assigned Amount and any additions or reductions to the party's national registry, which records transactions in emission reduction units (ERUs) from joint implementation projects, certified emission reductions (CERs) from the Clean Development Mechanism, assigned amount units (AAUs) from emissions trading, and removal units (RMUs) from afforestation, reforestation and deforestation-related activities. If the total reported emissions of regulated greenhouse gases exceed the party's Assigned Amount, plus or minus any transactions in ERUs, AAUs, CERs and RMUs, the party is in non-compliance.

It is difficult to anticipate what aspects of this calculus will come into dispute in the context of any particular case. The main enforcement consequence that applies to a failure to reach the agreed emission target is the application of a multiplier that yields a stiffer penalty relative to the size of the excess tonnes of emissions. There will be incentives for a party to contend every aspect of this calculus. National inventories of emissions and national registries of transactions that are self-reported by each party are checked annually by ERTs, with adjustments and corrections proposed by the ERTs. If a party disagrees with an adjustment or correction proposed by the ERT, the disagreement is to be resolved by the Enforcement Branch. Therefore, in theory, each Annex I party should arrive at the end of the commitment period with an uncontestable inventory, which can be compared against an uncontestable registry.

But the simplicity of this calculus, which suggests a high degree of automaticity, and the limited degree of discretion in the Enforcement Branch's decision making, masks an exercise that will require considerable judgment and discretion, particularly with regard to the calculation, assessment and adjustment of national inventories of emissions (see Chapter 3 by Mitchell and Chapter 4 by Berntsen and associates). Assessments made by an ERT, and reviewed by the Enforcement Branch, will not be on the basis of end-of-pipe measurements, but rather will be based on whether a party has followed good practice in applying formulae that estimate and extrapolate emissions from input and output data.

The guidelines for adjustments indicate that they shall be applied by the ERT:

> only when inventory data submitted by parties included in Annex I are found to be incomplete and/or are prepared in a way that is not consistent with the *Revised 1996 IPCC Guidelines for National Greenhouse Gas Inventories* as elaborated by the IPCC good practice guidance and any good practice guidance adopted by the Conference of the Parties serving as the Meeting of the Parties to the Kyoto Protocol.

Good practice, in turn, is a subjective judgement of whether a party's greenhouse gas inventories are:

> accurate in the sense that they are systematically neither over- nor underestimated as far as can be judged, and that uncertainties are reduced as far as possible. Good practice covers choice of estimation methods appropriate to national circumstances, quality assurance and quality control at the national level, quantification of uncertainties, and data archiving and reporting to promote transparency.

The enforcement procedures concentrate both power and discretion in the ERTs and the Enforcement Branch. While any party can raise a question of implementation with regard to the compliance of any Annex I party with its target, it is anticipated that most enforcement procedures will be triggered by a question raised by an ERT. Even if a party does trigger the procedure directly against another party, it is likely that the reports of the ERT will be extremely influential in determining the outcome. As has been described, if a party contests the findings of an ERT, the disagreements will be resolved by the Enforcement Branch. In practice, this appears to mean that by the time a party reaches the final reckoning of its compliance, any interim objections it may have raised along the way will have already been resolved by the same body, the Enforcement Branch, that is now assessing the overall results.

In adversarial procedures, such as cases before the ICJ and (to a lesser extent) the WTO, due process is provided by balancing the arguments of the respondent against those of a claimant, through procedures moderated by rules relating to burdens of proof and admissibility of evidence. The Kyoto procedure is, however, essentially non-adversarial. Although the ERT will likely raise the initial question of implementation, its mandate precludes it from making any judgements as to non-compliance, and it does not have a role in prosecuting the case once the procedure has been triggered. This suggests that the party's main interlocutor will be the Enforcement Branch itself. This is very similar to the dynamic of the Montreal Protocol's Implementation Committee.

A party's case will first be weighed after a question of implementation is allocated to the Enforcement Branch. Any question of implementation regarding compliance with targets would be allocated to the Enforcement Branch. The Branch must then undertake a 'preliminary examination of the question before it' and *ensure* that the question is supported by sufficient information, whether the question is *de minimis* or ill-founded, and whether it is based on the provisions of the Protocol.

The Enforcement Branch must, therefore make a positive determination that the question does have a basis in fact and in law, suggesting that the burden of establishing a *prima facie* case lies with the ERT or the party that has raised the question. There is no guidance in this or other agreed text as to what constitutes 'sufficient information' or what would fall below a *de minimis* threshold of non-compliance. While a party concerned may comment on all information relevant to the question raised against it, it does not appear to have the opportunity to intervene until after the decision to proceed with the question has been made.

As has been described, the decision by the Enforcement Branch to allow a question of implementation to move forward will be governed by voting rules that require the Branch to make every effort to reach consensus. If efforts to reach consensus fail, the decisions are made by what is known as a double majority. Assuming all members of the branch are present and voting, any decision must be approved by three-quarters of the membership as a whole (at least eight of ten members) and a simple majority of the Annex I members (at least three of four Annex I members). Put another way, two Annex I members could, by preventing the Enforcement Branch from taking a positive decision on the *prima facie* validity of the question of implementation, prevent that question from moving forward. If Annex I members will be more sympathetic to Annex I parties (as many have assumed they will be) this voting rule, which is applicable to all of the Enforcement Branch decisions, seems to afford these parties a significant degree of procedural protection against unsubstantiated claims.

If a question of implementation regarding compliance with a party's target clears the preliminary examination, it will move forward into the general and specific procedures governing the Enforcement Branch. As has been described, the Annex I party is entitled to be represented before the Branch, to submit information to the Branch, and to have access to and provide its written comments on any information submitted by others. Other than this opportunity to comment on others' information, there are no evidentiary rules entitling the party concerned to challenge the admissibility or relevance of information that is submitted. Each finding of the Branch must, however, be accompanied by 'conclusions and reasons therefor', suggesting that the Enforcement Branch will have to provide a rational basis for why it accepted some information as valid, and why it rejected other information.

The decision of the Enforcement Branch with regard to a party's compliance with its target requires a positive determination by the branch. This will involve a determination of not only whether a party is in non-compliance, but also, presumably, could involve a determination of the degree, measured in number of tonnes of carbon equivalent, that the party is out of compliance. While there is no provision at this stage of the process for a decision that non-compliance is *de minimis*, because the size of the enforcement consequence is relative to the degree of non-compliance, a small amount of excess yields a minimal penalty. The presumption built into the text is that a party is in compliance, and the Branch's mandate is to determine 'whether a party included in Annex I is *not* in compliance with its target'. This decision, should consensus fail, will be governed by the double majority rule described above. Once again, two Annex I members (and any combination of three members) can block a determination of non-compliance.

The final due process protection offered to a party that has been found to have exceeded its assigned amount is the right to appeal. As has been noted, the grounds for an appeal are limited to decisions made by the Enforcement Branch in circumstances in which the party feels it has been 'denied due process'. The provision does not indicate by what standard of review the Enforcement Branch's decision will be reviewed, but the procedural requirement that it can only be overturned by a three-quarters majority of the COP suggests a very high degree of deference will be accorded to the Branch's judgements.

**Table 2.2 Enforcement consequences**

| If the Enforcement Branch has determined that a party is not in compliance with... | ...the Enforcement Branch shall apply the following consequence to that party: |
|---|---|
| Its quantified emission limitation or reduction commitment under Article 3, paragraph 1, of the Protocol | deduction from the party's assigned amount for the second commitment period of a number of tonnes equal to 1.3 times the amount in tonnes of excess emissions;<br>development of a Compliance Action Plan; and<br>suspension of the eligibility to make transfers under the emissions trading provisions of the Protocol, until the eligibility is reinstated. |
| The methodological and reporting requirements under Article 5, paragraphs 1 and 2, and Article 7, paragraphs 1 and 4, of the Protocol | declaration of non-compliance; and<br>development of a compliance plan. |
| The eligibility requirements for participation in the flexibility mechanisms under Articles 6, 12 and 17 of the Protocol | suspension of eligibility under relevant mechanism, until eligibility is reinstated. |

### *The legal character of the enforcement consequences*

As has been indicated, during the negotiations on the design of the compliance system, reaching agreement on the availability of enforcement consequences in response to non-compliance depended upon a tight and predictable relationship between identifiable categories of non-compliance, and the consequence that would be associated with that category of non-compliance. This relationship is reflected in the text on the mandate and on the consequences to be applied by the Enforcement Branch.

The Enforcement Branch shall be responsible for 'determining' whether an Annex I party is in non-compliance with the listed commitments. This corresponds with Article 18 of the Kyoto Protocol, which states that the relevant procedures and mechanisms shall 'determine and address' cases of non-compliance. The use of 'determine' suggests that the finding of the Enforcement Branch is final, unless overturned on appeal. But, were decision of the Enforcement Branch to be raised in the forum of another tribunal, such as the ICJ, the question could arise as to whether it would be considered *res judicata*. Although the Enforcement Branch has similarities with an international court, the decisions of the branch may not be considered to have the legal effects similar to those of an international court.[59] Furthermore, the fact that the decisions of the Enforcement Branch are final does not necessarily mean that the concomitant consequences are legally binding. The meaning of 'binding' in this respect will be further examined below.

If the Enforcement Branch determines that there has been a violation of the listed commitments, Section XV of Decision 24/CP.7 of the Marrakesh Accords defines which consequences shall be applied in relation to each of the three groups of commitments. By using the word 'shall', the Enforcement Branch has no discretion in selecting the consequences it may find most suitable. While this lack of discretionary powers may promote foreseeability and prevent abuse of powers, the disadvantage is obviously that there is little possibility to design the consequences appropriate to the circumstances of each individual case.

*Non-compliance with Article 5 (1) and (2) and Article 7 (1) and (4)*

Article 5 (1) of the Kyoto Protocol establishes an obligation for Annex I parties to have in place a national system for the 'estimation of anthropogenic emissions by sources and removal by sinks' of the greenhouse gases covered by the Protocol. The inventory shall be based on the methodologies and adjustments provided in accordance with Article 5 (2). Article 7 establishes that Annex I parties shall incorporate in their annual inventory of emission by sources and removal by sinks the 'necessary supplementary information for the purposes of ensuring compliance with Article 3, to be determined in accordance with paragraph 4 below'.

It is established that the Enforcement Branch shall, when it has determined that a party is in non-compliance with these provisions, issue a declaration of non-compliance and require that the party submit a plan that includes an analysis of the causes of non-compliance, measures that the party intends to implement to remedy the non-compliance, and a timetable for implementing such measures. The party shall also submit progress reports on the implementation of the plan (XV (1), (2) and (3)). It is, however, provided that the consequences shall be applied 'taking into account the cause, type, degree and frequency of the non-compliance of that party'. This is consistent with Article 18 of the Kyoto Protocol, which also refers to the 'cause, type, degree and frequency of non-compliance'. The Enforcement Branch is thus left with discretion, not to decide on other kinds of consequences, but to design the designated consequences to the case at hand. It may even be questioned to what extent the use of these consequences is mandatory at all. The possibility for the Enforcement Branch to 'at any time, refer a question of implementation to the Facilitative Branch for consideration' may be pointed out.

The Enforcement Branch's issuing of a declaration of non-compliance is a typical soft sanction, in the sense that its effect is the shaming of the relevant party. Such a finding should be considered binding, but not necessarily binding in the sense of the Kyoto Protocol, Article 18, requiring an amendment to the Protocol (see below). A declaration of non-compliance does not in itself establish new obligations on the party. While the requirement to develop a plan to remedy the non-compliance is also a relatively soft sanction, it has the legal effect of imposing a new obligation on the party. The same may be said for the requirement to submit progress reports.

*Non-compliance with the eligibility requirements under Articles 6, 12 and 17*

If the Enforcement Branch finds that an Annex I party does not meet one or more of the eligibility requirements of the flexibility requirements of Article 6, 12 and 17, it shall 'suspend the eligibility of that Party in accordance with the relevant provisions under those articles' (XV (4)). This must mean that all uses of the flexibility mechanisms are prevented, be they transfer or acquisition of quotas, or use of joint implementation or the CDM. However, the presumption must be that loss of eligibility only refers to the particular kind of flexibility mechanism with which eligibility requirements are not fulfilled. Unlike non-compliance with Article 5 (1) and (2) and Article 7 (1) and (4), there is no discretion to take into account 'cause, type, degree and frequency of non-compliance'. At the request of the party, the eligibility may be reinstated under the procedure in section X (2), but the decision lies with the Enforcement Branch.

The suspension of eligibility may be of considerable economic and political importance for the party concerned, and should thus be considered a hard sanction. The sanction is binding in the sense that such a decision prevents the party from being credited with transactions under the flexibility mechanisms in meeting its obligations under Article 3 (1). However, it will be further discussed below whether this consequence is binding in the sense that amendment under the Kyoto Protocol is required.

*Non-compliance with the emission commitment under Article 3 (1)*

As has been discussed, if the Enforcement Branch determines that the emissions of a party have exceeded the amount assigned to it in Article 3 (1) of the Kyoto Protocol, it shall declare that the party is in non-compliance with this provision, and shall apply the following consequences (XV (5)):

1 deduction from the party's assigned amount for the second commitment period of a number of tonnes equal to 1.3 times the amount in tonnes of excess emissions;
2 development of a Compliance Action Plan in accordance with certain requirements (paragraphs 6 and 7) in order to reachieve a status of non-compliance; and

3 suspension of the eligibility to make transfer as part of emissions trading under Article 17 of the Protocol, until the party is reinstated by the Enforcement Branch in accordance with the procedures in Section X (3) or (4).

The party shall submit a progress report on the implementation of the Compliance Action Plan on an annual basis (paragraph 7).

The soft, hard and/or binding character of the declaration on non-compliance, the development of a Compliance Action Plan and the submission of progress reports have been touched upon above. The new elements are the deduction of tonnes of emissions for subsequent years at a penalty rate, and suspension of the eligibility to make transfers under Article 17. It should be mentioned that the Marrakesh Accords distinguish between transfers and acquisitions of quotas.[60] Accordingly, parties are only prohibited from selling, not buying quotas in cases of non-compliance.

It has already been argued that suspensions of eligibility to make transfers reflect hard enforcement consequences. They are also binding in the sense that any tonnes transferred in violation of the suspension are not valid, and thus the party will in practice have nothing to sell. The deduction of tonnes is obviously also a hard sanction; the legal question is whether this and the other consequences should be considered binding in the sense that they may require an amendment of the Kyoto Protocol under its Article 18.

But let us first consider the appropriateness of using fixed consequences rather than leaving the choice to the Enforcement Branch. It has already been stated that fixed consequences have the benefit of promoting predictability and preventing abuse of powers. It may furthermore be argued that at a general level, the consequences chosen are well designed as reactions to the relevant violations. But the use of fixed consequences prevents the choice of consequences designed to the specifics of each individual case.

Given the composition and the due process guarantees of the Enforcement Branch, there is good reason to argue that this branch should have been entrusted with the discretion to choose the appropriate consequences, rather than establishing certain mandatory consequences. It may also be asked whether the use of fixed consequences is consistent with Article 18 of the Kyoto Protocol, requiring an 'indicative list' of consequences, 'taking into account the cause, type, degree and frequency of non-compliance'.

*The legal status of the enforcement consequences*

As has been described in detail in Chapter 1, delegations debated heatedly whether, and, if so, how, to provide the enforcement consequences with legally binding force. The legal character of these consequences was directly linked to the form by which they were to be adopted by the parties. This link is made by the terms of Article 18, which provide that 'any procedures and mechanisms under this Article entailing binding consequences shall be adopted by means of an amendment to this Protocol'. The link is enforced by basic principles of international law, which generally require that a state must consent, in advance, to any international rule that can be considered to be binding on it. Although, by ratifying the Protocol, each party consents generally to the responsibilities that arise from its breach, it is far from clear that these, as a matter of treaty or custom, would necessarily and automatically equate to the enforcement consequences included in the Marrakesh Accords.

Delegations that supported legally binding consequences tried to find a way around Article 18's amendment provision, as this would have led to complex ratification procedures, and to the possible delayed and uneven application of the compliance system to those parties that were slow or chose not to ratify the amendment. The Marrakesh Accords leave both the issue of the legal character of the consequences, and the form by which they will be adopted, unresolved, agreeing instead that it is the prerogative of the COP/MOP 'to decide on the legal form of the procedures and mechanisms relating to compliance'.

It was clear from the deliberations that those who opposed the adoption of legally binding consequences viewed the language in Article 18 as a procedural guarantee that no party can be bound by enforcement consequences without the opportunity to consent, through ratification, to any procedure that might lead to those consequences. If this position prevails at COP/MOP 1 the result may be a two-track approach, whereby the COP/MOP adopts the compliance system by decision, and simultaneously adopts an amendment that will be open for ratification. The compliance system would become immediately operational, but the risk would remain that at the end of the commitment period, an Annex I party might be found to be in non-compliance but to have not yet ratified the amendment.

The implications of this remaining legal ambiguity may be different for each of the enforcement consequences. It should first be stated that the determination of non-compliance, although being a legal determination and having a shaming effect, does not in itself entail any consequences. Hence, it is less relevant to its effectiveness if this consequence is considered binding.

The suspension of eligibility to participate in the mechanisms raises different issues. The availability of the use of flexibility mechanisms may, for example, be regarded as a privilege granted by the treaty organs rather than a right flowing from the treaty itself. It may be argued that the relevant organs of the climate change regime should have similar powers to those enjoyed by organs of IGOs, particularly so-called 'implied powers'. If so, it might be said that it is within the discretion of these organs whether or not to grant such privileges, and to suspend them if the necessary conditions are not fulfilled.[61] The requirements to adopt a Compliance Action Plan and submit progress reports on implementation may also be within the implied powers of the Kyoto Protocol's organs.

If assigned amounts for a second commitment period have been agreed by the time a penalty rate is applied to a party that has been found to have exceeded its assigned amount, the resulting removal of tonnes from that second assigned amount will be a new and real burden for a party. The ongoing integrity of the Kyoto regime depends upon that additional burden being viewed as a binding element of that second commitment period cap. Although the deduction in tonnes from the second commitment period may be considered internally binding for organs established under the Protocol, including the COP/MOP, parties that have withheld their consent to Marrakesh consequences may have a legal basis for arguing that they are not bound by these deductions.[62] The language of Article 18 may provide a last bulwark of due process for those parties that have felt hard done by the Enforcement Branch.

## Conclusions

We have seen that the Marrakesh Accords – despite political controversy – have included enforcement consequences that should be considered to be of a hard character in cases of non-compliance. While also including soft elements, the Accords combine the facilitative approach and the enforcement approach to the implementation of international commitments. The inclusion of hard enforcement consequences may be explained by the stringency of the commitments, both in economic and political terms.

Whereas the determination by the Enforcement Branch of a party's non-compliance should be considered binding, an important unresolved issue is the legal status of the enforcement consequences. Suspension of eligibility to use the flexibility mechanisms, as well as requirements to adopt a Compliance Action Plan and to submit progress reports, may be considered to be within the competence of the Enforcement Branch. It may, however, be argued that Article 18 of the Kyoto Protocol means that binding deductions of assigned amounts at a penalty rate require an amendment of the Protocol. This may be a challenge to the effectiveness of the enforcement regime. It may also be asked whether the chosen consequences will work as effective deterrents.

The Marrakesh Accords contain several safeguards to ensure due process when consequences in the form of hard enforcement consequences are being considered. The impartiality of the ERTs and the Enforcement Branch, the qualifications of the ERTs and the Enforcement Branch with respect to factual and legal information, the opportunity of the concerned party to make its voice heard, the information upon which decisions shall be made, and the fixed consequences are all well designed to give credibility and legitimacy. The Secretariat is preserved as a neutral organ, while the political organ in the form of the COP/MOP has a very limited role to play. The process is multilateral in its approach: regard is thus being had to the multilateral character of implementation of the Protocol.

We have many treaty-based organs whose task it is to control implementation of multilateral commitments, be they in other MEAs, human rights treaties or disarmament treaties. But there are probably no other treaties where such organs may decide on enforcement consequences as hard as those in the Kyoto Protocol. There is also no other treaty – except in treaties establishing international courts – where comparable requirements regarding due process may be found. In this sense, the Kyoto Protocol with its Marrakesh Accords is a unique example of a treaty combining a multilateral approach, due process and hard enforcement consequences. In fact, the Protocol system seems to be the only one in international law to apply consequences of a penal character. Maybe this is a new feature of the law of a globalized world.

## Notes

1 Malmö Ministerial Declaration, 2000.

2 UNEP Governing Council Decision 21/27, *Compliance with and enforcement of multilateral environmental agreements*, 2001. UNEP's Programme for the Development and Periodic Review of Environmental Law (Montevideo III) as adopted in the meeting of the

Senior Government Officials Experts in Environmental Law, 23–27 October 2000, also contains a section on Implementation, Compliance and Enforcement.

3 UNEP Governing Council Decision SS.VII/4. *Compliance with and enforcement of multilateral environmental agreements*, Report of The Governing Council on the Work of its Seventh Special Session/Global Ministerial Environment Forum, 13–15 February 2002.

4 The Declaration from the Fifth Ministerial Conference Environment for Europe, Kiev, 21–23 May 2003 also 'stress[es] that greater emphasis should be placed on compliance with and national implementation of' MEAs (ECE/CEP/94/Rev 1 para 18) and the Conference endorsed the Guidelines for Strengthening Compliance with and Implementation of MEAs in the UNECE Region 'as an important tool to strengthen compliance with and implementation of regional environmental conventions and protocols' (para 43).

5 1997 Montreal Protocol on Substances that Deplete the Ozone Layer (1522 UNTS 293), 1979 UN ECE Convention on Long Range Transboundary Air Pollution (TIAS No 10, 541, 1302 UNTS 217) and 1998 Convention on Access to Information, Public Participation in Decision-making and Access to Justice in Environmental Matters (38 ILM 517 (1999)).

6 See, from a large literature Cameron, Werksman and Roderick, 1996; Széll, 1997; Weiss and Jacobson, 1998; Victor, Raustiala and Skolnikoff, 1998; Wolfrum, 1998; and Fitzmaurice and Redgwell, 2000.

7 See, about the managerial and enforcement models, Chayes and Handler Chayes, 1995; Danish, 1996–97; Raustiala and Slaughter, 2002; and Brunnée and Toope, forthcoming. It has been pointed out that '[t]he 1987 Montreal Protocol, the 1994 Second Sulphur Protocol and the 1997 Kyoto Protocol suggest a new cycle of strengthening compliance systems gradually, and in step with the strengthening of commitments' (Fitzmaurice and Redgwell, 2000, p43). Wang and Wiser, 2002, p184, claim that 'the strictness and comprehensiveness of a compliance regime under an MEA depends to a significant extent on the nature of the commitment embodied in the agreement'.

8 Marauhn, 1996, p696 points to 'the need to develop a procedural structure and procedural safeguards for compliance control'.

9 The parties also failed to agree upon the composition of the Multilateral Consultative Committee to be established under Art 13 of the FCCC (Oberthür and Ott, 1999, p213).

10 Kyoto Protocol, Art 8(3).

11 Kyoto Protocol, Art 8(2).

12 Marrakesh Accords, Decision 23/CP.7 Guidelines for review under Art 8 of the Kyoto Protocol. Annex (FCCC/CP/2001/13/Add.3).

13 Id, para 23.

14 Id, para 21.

15 Id para 25.

16 Id, para 32.

17 Marauhn, 1996, p712, empasizes the need to 'distinguish between factual and legal evaluation'.

18 Marrakesh Accords, Decision 23/CP.7 Guidelines for review under Art 8 of the Kyoto Protocol. Annex, para 4.

19 Id, para 5.

20 Id, para 6.

21 Id, para 48 (e).

22 Decision 24/CP.7 Procedures and mechanisms relating to compliance under the Kyoto Protocol, Annex, Sec II, paras 1, 2 and 3.

23 Id, Sec III.

24 Id, Secs XI and XIII.

25 Id, Sec VII, para 1.

26 Id, Sec II, para 3.
27 Id, Sec IV, para 1.
28 Id, Sec II, para 6.
29 Id, Sec II, para 6.
30 Id, Sec IV, para 3.
31 Id, Sec IV, para 4.
32 See also Wang and Wiser, 2002, p190, footnote 72.
33 Decision 24/CP.7, Annex, Sec IV, para 5 (a).
34 Id, Sec IV, para 5 (b).
35 Id, Sec IV, para 6 (a).
36 Id, Sec IV, para 6 (b) and (c).
37 Id, Sec II, para 9.
38 Id, Sec XIV (a)–(d).
39 Id, Sec II, para 3.
40 United Nations Framework Convention on Climate Change Draft Rules of Procedure of the Conference of the Parties and its Subsidiary Bodies (as applied) FCCC/CP/1996/2 22 May 1996, Rule 22.
41 Decision III/20 of the Montreal Protocol Meeting of the Parties, Composition of the Implementation Committee.
42 WTO Charter, Dispute Settlement Understanding, Art 17.
43 If an expert from Japan, which is the only Annex I member of the Asia Group, is elected to serve in this position, the totals would shift accordingly.
44 Decision 24/CP.7, Annex, Sec II, para 6.
45 Id, Sec II, para 6.
46 Id, Sec V, para 3.
47 Id, Sec V, para 4.
48 Art 2 of the Statute of the International Court of Justice.
49 Annex II of the Report of the tenth Meeting of the Parties to the Montreal Protocol.
50 Decision 24/CP.7, Annex, Sec VIII, para 3.
51 Id, Sec VIII, para 4.
52 Id, Sec VIII, para 5.
53 WTO Dispute Settlement Understanding, Art 8.3.
54 Decision 24/CP.7, Annex, Sec II, para 9.
55 Id, Sec XI, para 1.
56 Id, Sec XI, para 3.
57 Id, Sec XIII.
58 Art 3, paragraphs 7 and 8.
59 See also Fitzmaurice and Redgwell, 2000, p48.
60 See eg Decision 18/CP.7 and Paragraph 2 in the Annex to the proposed COP/MOP decision on Art 17.
61 See Churchill and Ulfstein, 2000, p647.
62 See also Brunnée, 2003, p278 on the practical effect of regarding the consequences as 'binding' in the Kyoto Protocol.

## References

Brunnée, J. (2003) 'The Kyoto Protocol: Testing Ground for Compliance Theories?', *Zeitschrift für ausländisches öffentliches Recht und Völkerrecht*, vol 63, no 2, 255–81

Brunnée, J. and Toope, S. J. (forthcoming), 'Persuasion and Enforcement: Explaining Compliance with International Law', *Finnish Yearbook of International Law*, Ius Gentium Association, Helsinki, Finland

Cameron, J., Werksman, J. and Roderick, P. (eds 1996) *Improving Compliance with International Environmental Law*, Earthscan, London

Chayes, A. and Handler Chayes, A. (1995) *The New Sovereignty. Compliance with International Regulatory Agreements*, Harvard University Press, Cambridge, MA

Churchill, R. R. and Ulfstein, G. (2000) 'Autonomous Institutional Arrangements in Multilateral Environmental Agreements: A Little-Noticed Phenomenon in International Law', *American Journal of International Law*, vol 94, no 4, 623–60

Danish, K. (1996–97) book review of Chayes and Handler Chayes, 'The New Sovereignty', *Virginia Journal of International Law*, vol 37, 789.

Fitzmaurice, M. A. and Redgwell, C. (2000) 'Environmental Non-Compliance Procedures and International Law', in *Netherlands Yearbook of International Law*, vol XXXI, T.M.C. Asser Press, The Hague, 33–65

Marauhn, T. (1996) 'Towards a Procedural Law of Compliance Control in International Environmental Relations', *Zeitschrift für Ausländisches Öffentliches Recht und Völkerrecht*, vol 56, 696–732

Oberthür, S. and Ott, H. E. (1999) *The Kyoto Protocol: International Climate Policy for the 21st Century*, Springer, Berlin

Raustiala, K. and Slaughter, A.-M. (2002) 'International Law, International Relations and Compliance' in Carlsnaes, T., Risse, T. and Simmons, B. A. (eds), *Handbook of International Relations*, Sage Publications, London

Széll, P. (1997) 'Compliance Regimes for Multilateral Environmental Agreements: A Progress Report', *Environmental Policy*, vol 27, 304

Victor, D. G., Raustiala, K. and Skolnikoff, E. B. (eds 1998) *The Implementation and Effectiveness of International Environmental Commitments: Theory and Practice*, MIT Press, Cambridge, MA

Wang, X. and Wiser, G. (2002) 'The Implementation and Compliance Regimes under the Climate Change Convention and its Kyoto Protocol', *Review of European Community & International Environmental Law*, vol 11, 181–98

Weiss, E. B. and Jacobson, H. K. (eds 1998) *Engaging Countries: Strengthening Compliance with International Environmental Accords*, MIT Press, Cambridge, MA

Wolfrum, R. (1998) 'Means of Ensuring Compliance with and Enforcement of International Environmental Law', *Recueil des Cours*, vol 272, 9–154

# Part II

# Challenges to Effective Operation of the Compliance Regime

Chapter 3

# Flexibility, Compliance and Norm Development in the Climate Regime[1]

Ronald B. Mitchell

## Introduction

In the Framework Convention on Climate Change (FCCC) of 1992, the world's nations aspired to stabilize 'greenhouse gas (GHG) concentrations in the atmosphere at a level that would prevent dangerous anthropogenic interference with the climate system'.[2] To achieve that lofty goal, the regime will need to create, over the next century, a broadly-held and abiding norm among governments and within global society that 'appropriate' behaviour requires significant and consistent efforts to reduce emissions of GHGs.

Other chapters in this volume focus on the relatively direct and immediate effects of the climate regime's existing compliance and enforcement mechanisms (or alternatives thereto), evaluating whether, how and under what conditions these mechanisms will influence the climate-altering behaviours of states and substate actors. In this chapter, I look at these mechanisms in light of their potential to have longer-term impacts that may, at present, not be as visible. I argue that initial compliance rates are likely to be quite high but that some, and perhaps much, of this compliance will be achieved without significant behavioural change in industrialized countries, economies in transition or developing countries. Taken together, the definitions of compliance, the methodologies for baselining, the flexibility states have in fulfilling their obligations, the availability of cheap emissions credits for violators to come into compliance, and the regime's general strategy of responding to non-compliance by facilitation rather than enforcement, will make it relatively easy and cheap to fulfil obligations. Initially, therefore, compliance rates are likely to be quite high even with little if any use of the regime's enforcement mechanisms. However, these high compliance rates are not likely to contribute significantly to averting climate change, at least in the short term. Such initial compliance, even if 'empty', may nonetheless help establish a strong international norm that countries that fail to take action to reduce GHG emissions are acting inappropriately. The very ease of complying makes it more difficult for a country to sustain an argument that not complying is appropriate: if compliance is easy for the country complying but non-compliance is expensive to the environment and other countries, then non-compliance is likely to be increasingly viewed as deserving of social opprobrium.

Were such a norm to develop, it could, over the longer term, contribute significantly to climate progress by convincing states that action to prevent climate change is the right thing to do even when it is costly.

At present, governments and individuals act in ways that contribute to climate change because of the strength of material incentives for such behaviours and the weakness of norms framing such behaviours as inappropriate. Existing social structures create both a logic of consequences and a logic of appropriateness that lead most governments and individuals to make little if any effort to avert climate change (March and Olsen, 1998). In a logic of consequences, 'action by individuals, organizations, or states is driven by calculation of its consequences as measured against prior preferences' (March and Olsen, 1998, pp949–50). Currently, most governments and individuals consider the costs of acting to avert climate change to exceed the benefits of doing so. There are, of course, a few 'no regrets' policies and behaviours regarding climate change in which the benefits accruing to those adopting them may be large enough to convince them to take action; new scientific insights may increase the number of such policies and behaviours. However, for the foreseeable future, the benefits of reducing or sequestering emissions or other climate-protecting actions will be uncertain, will occur in the future and will accrue to others, while the costs are clear, present and borne by those taking such actions, making such actions unattractive for those making decisions based in a logic of consequences. International institutions, including the climate regime, are systematically weak and have limited resources to alter costs and benefits enough to alter such decisions.

Given this, the climate regime's long-term success more likely depends on its ability, over the next century, to establish a new logic of appropriateness about acts that contribute to climate change. In a logic of appropriateness, actions are guided by norms that link particular identities or roles with particular behaviours in particular situations (March and Olsen, 1998, p951). Actors ask '"what kind of situation is this?" and "what am I supposed to do now?" rather than "how do I get what I want?"'(Finnemore and Sikkink, 1998, p914; see also the Introduction to this book). In this logic, actions are responses to internalized norms, externalized social expectations regarding appropriate behaviour in a situation, or beliefs regarding the behaviours one must engage in to acquire or maintain a certain identity. Actions that cause climate change, and the failure to take actions to avert it, generally have not yet been framed as illegitimate, reprehensible or otherwise inappropriate. Climate regime institutions and processes must foster economic, political and social changes that make behaviours that contribute to climate change appear increasingly inappropriate and convince more and more actors to accept that acts that are costly from a material logic of consequences perspective are, nevertheless, worth undertaking because they are the right thing to do.

Ultimate regime success at altering actors' behaviours depends less on instrumental manipulation of incentives than on deeper transformations of the goals they embrace and the behavioural norms they accept. Non-authoritarian governance – the only type possible internationally – works only if those being governed view its proscriptions and prescriptions as legitimate and appropriate constraints on their behaviour. Behavioural change will require states and their policy makers to accept either an internalized or externalized behavioural norm, either believing themselves that actions to avert climate change are appropriate obligations or believing that other states and their own polities view such actions as appropriate obligations. Experience with human rights norms suggests that, while it may take decades, relatively weak international institutions can induce such normative shifts that eventually convince even initially-resistant governments to bring their behaviours in line with those considered necessary to be accepted as full members of the community of nations (Keck and Sikkink, 1998).

I begin by delineating how compliance is defined and the role of baselines in both country-level and project-level compliance under the FCCC, as developed up to and including the provisions of the Marrakesh Accords. I then go on to examine how the provisions for monitoring and verification, review, facilitating compliance and enforcement are likely to operate. I examine how those compliance and enforcement mechanisms are likely to influence the behaviour of states and non-state actors in the short term through a logic of consequences, and how they may have deeper and more substantial influences on their behaviour over the longer term through a logic of appropriateness.

## Obligations under the FCCC

Under the 1997 Kyoto Protocol to the FCCC, many industrialized countries and economies in transition listed in the Convention's Annex I have undertaken quantified emission limitation or reduction commitments (QELRCs) delineated in the Protocol's Annex B, which are intended to achieve, in the aggregate, a 5 per cent reduction in these countries' emissions from 1990 levels by 2008–12.[3] The Marrakesh Accords of November 2001 elaborated compliance definitions and mechanisms related to these commitments.

Were each Annex B country required to achieve its QELRC target through domestic action, the behavioural changes required would have been both costly and inefficient. The environmental benefits of reducing or sequestering a ton of carbon dioxide are independent of where the reduction occurs, but the corresponding costs of this reduction or sequestration vary significantly across countries. Requiring countries to meet their commitments domestically would make achieving a given reduction more costly and, therefore, would reduce the total reductions politically and practically achieveable. To address this, the Marrakesh Accords granted states considerable flexibility in meeting their commitments. This flexibility involves creating an emissions accounting and trading system in which countries are granted emission units corresponding to their Annex B commitments, can buy or sell additional emission units under specified conditions, and will be considered compliant if the total number of emission units they possess exceeds their actual emissions. The Marrakesh Accords established four types of emission units and usage rules, as follows:

1 assigned amount units (AAUs), which correspond to a country's Annex B commitment, and which another country can acquire through emissions trading from other Annex I countries (Article 17).

2 Emission reduction units (ERUs), which countries can acquire from other Annex I countries through joint implementation (JI) projects (Article 6).
3 Certified emission reductions (CERs), which countries can acquire from Clean Development Mechanism (CDM) projects undertaken in non-Annex I countries (Article 12).
4 Removal units (RMUs), which countries can acquire through activities in Annex I countries that remove GHGs from the atmosphere.

Countries were also allowed to 'bank' emission units in excess of their first commitment period targets and use them in future commitment periods. The banking of AAUs was not limited; the banking of ERUs and CERs was limited to 2.5 per cent of a country's original AAUs; and the banking of RMUs was prohibited. Since these provisions made emission reduction obligations dynamic, the parties created a transaction log system to track exchanges between countries and across periods and thereby provide clarity regarding each country's target at the end of the first commitment period.

The Marrakesh Accords established several other provisions. Parties were prevented from using these mechanisms unless they were compliant with mechanism-specific requirements as well as more general methodological and reporting requirements.[4] Countries were provided a period, called a 'true-up'period, during which they can acquire enough AAUs, ERUs, CERs, and/or RMUs to fulfill their commitments that extends for 100 days after completion of the expert review process for the last year of the first commitment period, i.e. 2012.[5] A version of 'buyer liability' for the validity of permits was created by requiring countries to maintain a commitment period reserve of 90 per cent of their assigned amounts, calculated from the QELRCs reflected in Annex B, to reduce the risk of countries overselling emission units and failing to meet their targets.[6] Developing countries were authorized to undertake unilateral CDM projects and market the resulting emission credits (Pew Center on Global Climate Change, 2001).[7] In addition, a CDM Executive Board was created to develop rules governing the operation of, and verification and accounting of credits from, CDM projects, including accrediting operational entities to evaluate projects and assign project credits.

The Marrakesh Accords established a Compliance Committee, consisting of a Facilitative and an Enforcement Branch. The Facilitative Branch is tasked to promote compliance by all parties with their commitments through appropriate recommendations, advice and/or technical and financial assistance.[8] The Enforcement Branch is tasked with determining the compliance of Annex I parties with their Protocol commitments, and applying consequences that can include – depending on the commitment breached – a declaration of non-compliance, submission of a plan for coming into compliance, suspension of eligibility to participate in the flexibility mechanisms, and the deduction of 100 per cent of excess emissions in the first commitment period, plus a 30 per cent penalty, from a party's assigned amount in the second commitment period.[9]

## *The role of baselines*

Both country baselines and project baselines influence compliance dynamics within the climate regime. Country baselines are estimates of a country's net GHG emissions during the first commitment period (2008–12), had it followed a business as usual scenario and not changed its policies or actions in response to the regime. Project baselines are estimates of net GHG emissions from a specified area over a specified time had a JI, CDM or RMU project not been undertaken.

Since Annex B targets translate into absolute emission levels, country baselines do not enter into legal determinations of Article 3.1 compliance, which depends only on a country having valid AAUs, ERUs, CERs and RMUs in excess of their actual emissions. However, country baselines are central to how difficult and costly it will be for a country to comply, and therefore how likely it is that countries will. The distance between a country's business as usual baseline and its target corresponds with the degree of behavioural change a country must make (or the amount it must spend on credits) if it seeks to comply. Because GHG emissions tend to track economic growth, declining economic conditions will make it easier or cheaper to comply, while economic booms will make it more difficult or costly.

Country baselines, even when not actually known, are likely to be central determinants of the cost of emission units in whatever emissions trading market develops. Annex B commitments ensure that at least some countries, most notably Russia and the Ukraine, can, without taking any climate-protecting action, meet their Kyoto targets and still have 'hot air' credits to sell. Most other countries will have to estimate their emissions relative to their targets and either introduce policies and programmes to meet their commitments domestically, initiate JI or CDM projects abroad, or acquire AAUs, ERUs, CERs or RMUs from other countries in the emissions market. The decision of which action to take will depend, of course, on the relative cost per ton of carbon dioxide ($CO_2$)-equivalent of those alternatives.

The supply of, and demand for, hot air credits will influence the price of all emission units and hence the cost of compliance. Because there is no real cost of producing hot air units, Russia, Ukraine and any other countries that can confidently predict business as usual emissions below their commitments can sell those units at very low prices. Those countries will have incentives to charge as much as they can, but can undercut (and therefore drive down the price of) any other credits which those countries purchasing credits perceive as equivalent. The regime has sought to foster an emissions trading market in which AAUs, ERUs, CERs and RMUs are equivalent. Assuming the US neither ratifies Kyoto nor enters the emissions market during the first commitment period, estimates suggest that the 'overall demand for emission rights is likely to be lower than the supply of "hot air" from Russia and Ukraine [and so] the world market price will be very low' (Michaelowa, 2001, pVI). That said, two factors may create a market that differentiates hot air credits from others. First, given the buyer liability created by the 90 per cent commitment period reserve requirement, if uncertainty regarding the validity of hot air credits is greater than uncertainty associated with other credits, those credits will be priced differently (Victor, 2001a; Victor, 2001b). Second, some European states may restrict themselves in ways that go beyond the regime, including capping the amount of hot air credits they will allow themselves to apply towards their targets (David G. Victor, personal communication, 9 August 2002). Despite these caveats, hot air credits are likely to reduce significantly the attractiveness and price of project-based credits.

Project baselines pose problems that have more direct impacts on compliance, particularly the determination of non-compliance. Assessing the credits that a JI project (ERUs), CDM project (CERs) or sequestration project (RMUs) should receive requires very specific baselining. Credit assignment requires comparing emissions produced by the project to a baseline scenario 'that reasonably represents the anthropogenic emissions by sources or anthropogenic removals by sinks of GHGs that would occur in the absence of the proposed project'.[10] This task is analytically and empirically complex. The CDM Executive Board must approve methods for defining project boundaries and baselines, develop monitoring and evaluating procedures and accredit operational entities charged with those tasks, and maintain the registry of CDM projects and credits. Operational entities will approve the baseline and monitoring plans included in project design documents.[11] That will require clear definitions of project boundaries, i.e. the activities, area and time period to be considered part of the project, so that actual emissions for those activities, that area and that time period can be monitored and counterfactual emissions estimated. Defining projects and monitoring emissions are difficult but necessary elements in identifying reductions caused by the project, and therefore in determining the credits the project should receive.

Baselining engages 'the fundamental problem of causal inference' (King et al, 1994), the fact that we can never truly know what would have happened otherwise, and therefore can never be sure how much a given project reduced emissions. Any project has myriad, equally plausible counterfactual scenarios, each of which implies very different emission credits. Both overestimating and underestimating credit levels reduces the regime's ability to encourage emission reductions. The incentive to buy credits by funding JI, CDM and sink projects depends on the usefulness of those credits in fulfilling climate regime obligations at less cost than through emission projects undertaken at home. Excessively conservative baselines underestimate credits, reducing the incentives to fund such projects. Excessively liberal baselines overestimate credits, increasing participation in projects that are given more credits than they actually produce. Actors will have incentives for strategic behaviour aimed at establishing higher baseline emissions and therefore greater credit for a project. These and many other problems have been recognized and discussed in efforts to design systems to estimate counterfactual emissions, measure actual emissions and compare the two, in order to identify ways that mitigate, even if they do not eliminate, these problems. In practical terms, baselines and monitoring procedures that are flawed from theoretical perspectives may nonetheless be accurate enough that project participants, the regime secretariat and other contracting parties will accept them as appropriate bases for granting credits.

## Compliance mechanisms and procedures

The Marrakesh Accords provided interim closure in developing procedures to encourage compliance in the short term and maximize emission reductions in the longer term. These procedures involve verifying and reviewing behaviours and emissions, facilitating compliance prior to the end of the first commitment period, sanctioning non-compliance with inventory and reporting requirements prior to the end of the first commitment period, and sanctioning non-compliance with targets after the end of the first commitment period.

## *Monitoring and verification*

The climate regime's obligations adopt an economic efficiency (rather than 'command and control') approach, which defines compliance in terms of results states must achieve rather than actions they must take. The first step in evaluating compliance involves monitoring and verification. Collecting extensive and high quality behavioural and environmental information is important not only to validate whether actors deserve credit for reductions, but to learn more about human impacts on the climate and how best to reduce those impacts.

Under the climate regime, countries are required to develop national emission inventory systems, use those systems to provide annual inventories, and have both the system and the annual inventories verified by Expert Review Teams (see Chapter 2 by Ulfstein and Werksman and Chapter 4 by Berntsen and associates). Nominally, compliance requires only monitoring 'anthropogenic emissions by sources and removal by sinks'. In practice, however, major components of most countries' inventories will not be based on direct measurements of gases. Instead, they will estimate emissions and removals by applying emission rates to measurements of relevant activities. Methane emission levels may be based on counts of livestock by type multiplied by emissions per livestock type, and hectares of rice cultivation multiplied by emissions per hectare. Transport-related $CO_2$ emission levels seem likely to be based on consumption of different fuels multiplied by fuel-specific emission factors. Regardless of whether emissions are based on such calculations or direct measurements, the broader goals of the regime require monitoring both behaviours and emissions to better understand how variation in emissions depends on the type of, and conditions under which, policies and activities are undertaken.

The range of behaviours that emit or sequester GHGs precludes making general claims about the ease of monitoring (Morlot, 1998). In some cases, such as power plant emissions, relevant activities (amount of coal or oil burned) and environmental results (amount of $CO_2$ emitted) are relatively easy to monitor. In others, such as deforestation, relevant behaviours may be relatively easy to monitor (satellite surveillance of net changes in forest cover) but it may be more difficult to identify corresponding emissions because of problems in modelling carbon release. In yet other cases, even relevant behaviours may be difficult to monitor, as with determining the number of methane-producing livestock being grazed or levels of GHG-emitting military activities that governments have incentives to keep secret. The climate regime will need to develop mechanisms for evaluating, providing feedback on and making recommendations or even requirements for the models states use in estimating emissions from behavioural measurements.

## *Review*

Evaluating national-level compliance requires comparing the emissions information countries provide, however created, to the valid AAUs, ERUs, CERs and RMUs they hold. Since states will know what this review process will entail before submitting their emission inventories, the reviews themselves seem unlikely to 'catch' states in violation of their commitments (Victor et al, 1998b). States seem likely to enter the review stage using one of four strategies: they will have brought themselves into compliance by ensuring they have requisite emission credits; they will have constructed inventories that successfully misrepresent their actual emissions so they appear to be in compliance; they will acknowledge that they are out of compliance and seek to receive assistance in coming into compliance; or they will acknowledge that they are out of compliance and decide to accept or ignore whatever formal or informal sanctions may be imposed. Misrepresentation seems the least probable of these outcomes.

Thus, implementation reviews seem unlikely to uncover non-compliance. They may nonetheless contribute to the compliance process. They may identify why a particular state failed to comply and thereby identify more general factors that hinder or facilitate compliance. The justifications and explanations of a state's non-compliance and the evaluation of those justifications by other parties are likely to contribute to learning by the regime and other states about which policies are effective at reducing or sequestering emissions and under what conditions. Reviews may entail discussions about whether national governments, local governments, private actors, unexpected economic shocks or natural forces are responsible for any shortfalls; whether shortfalls reflect failures of policy or implementation; and whether intention, incapacity or inadvertence was the major cause of the shortfall. Review procedures will also need to identify or develop methodologies for addressing the many cases in which monitoring efforts face inherent difficulties in identifying which actors, if any, were responsible for a shortfall. Thus, it may be easy to identify those responsible for high emission rates from the power sector but far more difficult to identify those responsible for high emission rates due to deforestation or methane production.

In contrast to assessments of national-level compliance, project-level compliance explicitly requires reviewing both behaviours and emissions. Reductions or sequestration can only be shown to be additional if actual emissions, project-related behaviours and non-project factors (both human behaviours and environmental conditions) that influence emission/sequestration levels are observed or estimated. Projects are likely to be of four types: 'successes', in which participants' actions reduced or sequestered emissions; 'good faith efforts', in which participants' actions would have reduced or sequestered emissions were it not for other factors; 'coincidental compliers', in which reductions or sequestration occurred in spite of participants' actions or failures to act; and 'failures', in which participants' actions or failures to act caused the project not to produce reductions or sequestration. High-quality baselining systems should distinguish most, if not all, successes from coincidental compliers, so long as they identify changes in factors that influence project performance but change after initial baselines are agreed upon, such as economic growth rates or weather patterns. In a world in which other factors are never held constant, this is no easy task. It requires large amounts of accurate information about a large range of human behaviours and environmental conditions, as well as methodologies for converting that information into politically compelling assessments of which actors should be rewarded and which should be sanctioned.

The regime clearly seeks to build a transparent and accurate system of international monitoring, reporting, verification and review of emissions and behaviour. Its success depends not only on the initial structure and methodologies of those systems, but also on the incentives created to improve those systems over time. An important tension exists, however, between the desire for an accurate reporting system and the use of information from that system in determinations of non-compliance and application of sanctions by the Enforcement Branch of the Compliance Committee. States and non-state participants in JI, CDM and RMU projects know that the reductions or sequestrations they are credited with depend on actual emissions reported. They therefore have incentives to provide or withhold information, promote or oppose certain methodologies, and argue and negotiate over baselines and emissions in ways that maximize their reductions or sequestrations. How strong those incentives are depends on the response those actors expect for non-compliance and whatever supports and rewards are given for providing accurate information, especially if that information identifies non-compliance.

## *Facilitating compliance*

The regime's creation of a Facilitative Branch under the Compliance Committee represents only one element of a broader strategy based on facilitating compliance rather than deterring non-compliance. The regime relies largely on 'compliance management' while coupling it with some elements of an 'enforcement' approach (Chayes and Chayes, 1995; Chayes et al, 1995; Downs et al, 1996). The decision at Kyoto to allow states to choose their own targets, including ones that allowed for increases over 1990 levels, was a political necessity but does facilitate compliance. The flexibility mechanisms of emissions trading, JI and CDM projects, the use of sinks and RMUs, and the ability to bank emission units further increase the ways states can meet their targets while reducing the costs of doing so. Perhaps the largest provision facilitating compliance involves the 'additional period for fulfilling commitments'. This 'true-up' period allows states to acquire AAUs, ERUs, CERs and RMUs up until 100 days after completion of the expert review of inventories for the commitment period, a deadline that translates into over two years in which states can make up any shortfall (Michaelowa, 2001).[12] These provisions collectively make it much easier for states to comply, in part by reducing the need for behavioural change.

When states seem unlikely to comply, the regime has adopted a system 'to avoid confrontation, to be transparent' and to eschew sanctions in favour of cooperative measures.[13] The Facilitative Branch seeks to encourage Annex I parties to work with developing, non-Annex I parties in implementing their commitments, and to facilitate compliance by Annex I parties with building national inventory systems and meeting their targets. The Facilitative Branch can provide, or foster the provision by others of, advice, financial and technical assistance, and technology transfer and capacity building.[14] Developing countries involved in CDM projects or seeking to develop their inventory systems seem the most likely to make use of the Facilitative Branch, although Annex I countries that face unexpected difficulties may also do so. The success of such capacity building will depend on how many parties are committed to the regime's success, since only those struggling states that support the regime will request assistance and only those that support the regime will provide the financial and technological assistance envisioned. Unfortunately, historical experience suggests that funding, whether required of governments under the regime or volunteered by NGOs, is likely to fall short of that needed to create robust assistance programmes (Keohane and Levy, 1996; Victor and Salt, 1994, p15).

The regime has not delineated mechanisms to reward overcompliance and innovation. Since current emission reduction targets fall far short of what most scientists consider necessary to avert climate change, significant progress requires incentives for exceeding current targets and for undertaking risky projects that provide uncertain, but potentially large, reductions at low cost. At present, although the regime facilitates compliance, it has developed few mechanisms to reward those who go the next step, such as public awards, white lists, access to financial mechanisms or reduced project verification requirements.

## *Enforcement*

The regime also seeks to discourage non-compliance through mechanisms that include, but extend significantly beyond, the activities of the Enforcement Branch. At the project level, the incentives to discourage false credits will depend largely on the market's ability to distinguish (and discount the price of) false credits from those more likely to be assessed as valid. Governments may find the most salient reasons to comply involve the political criticism they receive from their own citizens, the media or other governments of failing to comply with a regime that has made compliance both easy and cheap.

For an emissions market to develop, there must be at least some governments committed to complying with their commitments and willing to promulgate corresponding domestic policies that pass these incentives on to subnational actors. These incentives to acquire credits will translate into a demand for credits only if those offering AAUs, ERUs, CERs or RMUs can convince potential buyers that those credits have value, i.e. that they are backed by real emission reductions or sequestration (Victor, 2001b, p22). Given the 90 per cent commitment period reserve, both corporations complying with domestic regulations and countries trying to fulfil national obligations must ensure that the credits they acquire will cover their emission obligations. Beyond the formal cancellation of credits, public and political sanctions for knowingly engaging in bogus transactions are also likely. Wary buyers are likely to require those undertaking projects to provide information reassuring them of credit validity prior to purchase, and to reward those that build reputations for high-quality projects and credits. This should, in turn, create a market in which false credits are distinguishable, harder to market and hence discouraged.

At the project level, sanctioning will require cancelling credits. The operational entities and the CDM Executive Board can be expected to reduce or revoke credits from projects that cannot demonstrate real reductions or sequestration corresponding to the project design. Despite the somewhat adversarial relationship of the verifying operational entities and the project participants, the former should be able to use the ability to provide or withhold credits to acquire data and information necessary to adequately validate credits. Project participants that do not want to monitor and report adequately will also build a reputation for not being forthcoming, making it more difficult for them to market those credits that they can get validated.

Operational entities will face difficulties in deciding whether to validate projects involving the coincidental compliers and good faith efforts mentioned earlier. Cancelling the credits of good faith efforts (e.g., a well-implemented reforestation project destroyed by a flood, drought or fire) has the virtue of ensuring that credits are backed by real reductions or sequestration, but punishes actions that reasonably could have been expected to produce such credits. The opposite problem is raised by projects of coincidental compliance (e.g., emission reductions from a plant that switched to a lower-emission fuel type which, after the project plan had been submitted, actually became cheaper than alternative fuels). Since the reductions were achieved by the planned activity, how should the fact that they were cost-free and would have occurred anyway (although this seemed unlikely when the project was approved) be taken into account? The CDM Executive Board and the operational entities will need to decide whether reductions from a project will be certified based on the actual behaviours of project participants or contracted behaviours. Even with buyer liability, buyers are likely to press for validation of credits in which contracted behaviour was carried out, regardless of actual emissions produced. As the regime develops procedures for allocating responsibility for project credit validation, it will need to consider how those procedures influence not only the actors involved in the case at hand but also the willingness of other actors considering undertaking such projects and trades in the future.

At the national level, the Enforcement Branch itself is entrusted with determining whether a state is in compliance, and taking action to urge it to come into compliance if it is not. Actions coming before the Enforcement Branch seem likely to be few, since the emissions trading market may develop in ways so that most states at risk for being sanctioned will either acquire credits during the true-up period or withdraw from the regime. Actions may come before the Enforcement Branch if concerns arise that a country's inventory system is susceptible to manipulation, but such issues will most likely be raised earlier in the Facilitative Branch, with the country seeking to respond to those concerns or rejecting treaty commitments outright. Despite the variety of sanctions available under the Marrakesh Accords, none seem likely to be used. Making participation in the flexibility mechanisms contingent on compliance with monitoring and reporting provisions will encourage states committed to the regime to develop strong inventory and reporting systems, but will have little influence on less committed actors. The regime's responses to non-compliance with targets appear somewhat harsher on paper but, like most sanctions in environmental agreements, seem unlikely to be used for several reasons. First, potential sanctioners will likely be deterred from imposing sanctions because of the political costs of accusing other states and of undermining a sense of common purpose in the regime (Axelrod and Keohane, 1986). Second, once the Enforcement Branch identifies a state as non-compliant it is mandated to demand the country provide a compliance plan, stop its sales of emissions credits, and make up 130 per cent of its shortfall in the next commitment period. Given the mandatory nature of these sanctions, the locus of contention will shift to the stage of calculating emissions and estimating (and re-estimating) baselines to avoid finding a country in non-compliance in the first place. Third, requiring states to make up 130 per cent of their shortfall in the second commitment period will lead those who expect to be found in non-compliance to make their emissions reduction targets for the second commitment period sufficiently small so as to not be burdensome.

If issues come before the Enforcement Branch, the problems of determining actual emission levels are likely to involve uncertainty rather than 'cheating'. Emission inventory systems are likely to be inaccurate and uncertain because of inherent measurement difficulties and a failure to dedicate adequate resources, rather than because of efforts to cheat. Thus, systems are more likely to lack data of sufficient quality to make a determination of non-compliance. Such data as the Enforcement Branch can acquire are likely to raise the question of what constitutes non-compliance. Since the 'actual emissions' numbers will almost certainly involve a mean estimate and corresponding confidence band, the question will be whether the mean, the 95 per cent confidence band or some other reference point should be used. If states seem likely to be found in non-compliance, they are likely to claim either that their higher emissions were due to large economic factors that prevented efforts they did make from being effective, or that they could not afford the credits necessary to come into compliance. Especially for states that took actions and expended resources that could reasonably have been expected to reduce emissions by a given amount, claims that they should be treated as if they complied are likely to prove compelling to other states and the Expert Review Teams.

## The regime's short-term influence: Altering the logic of consequences

Negotiators have developed obligations, compliance mechanisms and procedures to encourage climate-protecting activities that actors would not otherwise undertake. The mechanisms invoke a logic of consequences, seeking to alter behaviour by altering the incentives actors face. In the short term, the success of these mechanisms can be evaluated against two different, but equally important, standards. First, will they induce significant behavioural change? That is, will they lead to lower GHG emissions than would have occurred otherwise? Second, will they promote compliance? Will most states have sufficient valid emission units to cover their actual emissions? It is tempting to view the former standard as the only relevant standard. Indeed, since climate change will be slowed only if emissions are reduced or sequestration increased, compliance without significant behaviour change would seem to have little value. Yet, precisely because averting climate change will require decades, indeed centuries, of social management, the ultimate contribution of the regime's compliance mechanisms during the first commitment period may stem as much from their ability to establish norms against climate-changing behaviours as from their ability to alter behaviour. Compliance that reflects significant behavioural change is certainly preferable to compliance that does not. Yet, although the current regime seems likely to create more compliance than behaviour change, that compliance may well help establish climate protection norms that make it preferable to non-compliance, even non-compliance with more stringent rules.

In predicting the short-term influence of the regime, it seems appropriate to adopt a 'logic of consequences' model, assuming that actors choose behaviours by evaluating the relative costs and benefits of the alternatives available to them in a context in which their values and normative commitments can be treated as fixed. In the short term, the regime can increase the costs (or decrease the benefits) of non-compliance and non-participation while decreasing the costs (or increasing the benefits) of compliance. However, the underlying values of actors, which determine how those costs and benefits are weighed, are unlikely to change in the short term. Given this, it helps to estimate the responses to the regime by categorizing actors into four groups: committed, contingent, resistant and intransigent. Committed states are 'leader' states whose polities are already committed to a norm of taking action to avert climate change and whose choices are relatively insensitive to the costs and benefits the regime seeks to manipulate (Sprinz and Vaahtoranta, 1994). Contingent states are those that have not fully accepted such a norm but believe that action is probably warranted, depending both on the extent of action by other states and on their beliefs about the likelihood of, and harm from, climate change. Resistant states are those, the US being the most notable current example, that have rejected the norm that they should take action based on a view that the economic costs of action exceed the environmental benefits. Intransigent states are laggards or draggers who completely reject the norm that they should take action and whose behaviour has little to do with the costs and benefits the regime can manipulate. The regime was formed, and is largely populated, by an alliance of committed and contingent states. Some resistant states may have joined the regime, either because they sought the political benefits of membership and do not face significant obligations, or because their calculus of costs and benefits have changed since they joined and they have not yet withdrawn. Intransigent states can be assumed to have refused to join the regime from the outset.

How might we expect states in these groups to respond to the regime's compliance system in the first commitment period? Committed states are unilateralists whose choice to comply has little to do with the regime's flexibility mechanisms. They are nonetheless influenced by the effect of the regime's provisions on the costs of compliance. By reducing costs, the regime's flexibility encourages these states to take more domestic action than required, to make more behavioural change than required, and even, perhaps, to overcomply by reducing emissions below their targets. Lower costs allow these states to contribute more to averting climate change for whatever financial expenditure they were committed to making. These states are the ones most likely to provide financial and technical assistance, to fund JI and CDM projects in an effort to build capacity as well as commitment to the regime in other states, and to follow the spirit of the regime's rules that trading and acquisition of emission units 'be supplemental to domestic actions' (Art 6(1)(d) and 17).

The regime's flexibility is likely to have its largest effect on contingent states. Relative to what would have happened without these mechanisms, many more contingent states are likely to comply, since the mechanisms allow countries to meet many of their targets through trading in credits likely to be quite cheap (Michaelowa, 2001). Because contingent states want to comply so long as enough others do and it is not too costly, the true-up period is likely to prove particularly influential. During that period they will have considerably more information than at present regarding how many other states have complied, what they did to comply, how much credits cost and the risks of climate change. This information clarifies the material and social costs of complying and not complying. And, if the response of committed states is strong, the costs of complying are likely to be less and those of not complying greater, making contingent states more likely to comply. Unlike the committed states, however, they will be more likely to minimize their compliance costs, engaging in more 'empty' compliance. They are more likely to acquire credits or reduce emissions only as required and urge a liberal interpretation of the requirement that trading 'be supplemental to domestic action'. They are less likely to provide financial and technical assistance, develop accurate and reliable inventory systems, and ensure emission credits are backed by real reductions. In short, contingent states will tend to follow the letter rather than the spirit of the agreement.

Resistant states are unlikely to respond to the flexibility mechanisms in the short term. These states start by assuming that compliance will be costly and that non-participation and non-compliance will have few costs. Since most resistant states will not be parties, formal compliance mechanisms cannot be applied. Like contingent states, resistant states are sensitive to new information. Those that are parties will be likely to withdraw if their initial assumptions regarding the costs of compliance prove true. However, if the decision to not participate proves to entail significant domestic or international political costs and compliance proves to be relatively easy and cheap, these states may revise their decisions. They may join the regime or comply with it, or they may remain outside the regime while reducing emissions or even acquiring credits which, were they a member, would constitute movement toward compliance. Intransigent states, however, are as dedicated to opposing the regime as committed states are to supporting it. Although this does not seem like a large group at present, such a group may grow in future commitment periods as the incidence of the costs of averting climate change become clear. Thus, oil-exporting states may increasingly oppose the regime if most member states comply by reducing oil consumption.

This analysis suggests that, in the short term, the incorporation of flexibility into the regime is likely to lead to high levels of compliance without frequent use of the enforcement provisions that have been so carefully developed (Downs et al, 1996). These high compliance rates will reflect compliance coupled with significant behavioural change by committed states, compliance with far less behavioural change by contingent states, and the unwillingness of resistant and intransigent states to become or remain parties to the treaty and its requirements. The regime's flexibility mechanisms encourage greater behavioural changes by committed states, will encourage greater compliance by contingent states even if not accompanied by significant behavioural change, and may induce resistant states to reevaluate their position vis-à-vis the climate regime.

## The regime's long-term influence: Altering the logic of appropriateness

The climate regime's short-term influence is likely to be rather limited, since committed states would have taken much of the action anyway and contingent states are likely to comply through relatively minimal action and small expenditures to acquire credits. Yet these small near-term contributions to averting climate change may provide a foundation for long-term success at attracting participation, encouraging compliance and reducing GHG emissions. Flexible mechanisms that alter the logic of consequences in the short term may foster a broader social transformation in the logic of appropriateness surrounding efforts to protect the global climate. Changing behaviours by altering the logic of appropriateness, i.e. by altering the norms and values that inform the behaviours people and states choose, involves indirect and long-term processes. To the extent that climate regime components initiate social processes that provide sustained support for climate-protecting norms and values, their influence is likely to be positive and may well be considerable, however hard it may be to isolate analytically.

The contribution that flexibility mechanisms make to inducing most parties to comply may foster new norms regarding climate-related behaviours. Clearly, policies and programmes that produce real reductions in emissions and innovations in sequestration are important to such a shift. Yet, even where such efforts fail, the social signal sent by states engaging in such efforts reinforces the notion that taking action to avert climate change is the appropriate thing to do. Contingent states may find that compliance, which was initially engaged in because it was easy, over time becomes expected behaviour, deviation from which requires explanation (Young, 1992). When states comply with their targets, they simultaneously affirm the norm held by committed states that taking action to avert climate change is good while undermining claims by resistant states that it is excessively costly. The regime's flexibility also undermines claims of resistant states that the norm is illegitimate because it does not recognize that state's particular circumstances, or that the norm does not apply to its behaviour because it is not capable of complying. If incapacity is often used to defend against accusations of environmental non-compliance (Brown Weiss and Jacobson, 1998), the ability to comply by purchasing relatively cheap credits in a market created for this purpose removes such excuses. Regular meetings of the Conference of the Parties and subsidiary bodies provide opportunities for states to evaluate their progress and learn from others how to meet their commitments at lower cost. The Facilitative Branch provides a forum in which states can request help in meeting their commitments. All these provisions and institutions make it easier for a state to comply and, thereby, make it more difficult for it to claim it should not have to.

Perhaps the largest influence of the regime lies in its effect on how behaviours that contribute to climate change are framed. The Kyoto targets constitute rules 'around which actors' expectations converge' (Krasner, 1983). Although their nominal purpose is to create legal categories of compliant and non-compliant behaviours, their more important effect may be as foundations for social categories of identity. They become the basis for a broader social definition in which those that strive to reduce emissions below 1990 levels are considered 'green' and those that do not are considered 'brown'. Such social definitions need not strictly correspond to the regime's specific provisions. Governments that fall short of their Kyoto commitments may still be considered green if others believe they strived toward those targets. Equally importantly, resistant states may be considered brown whether or not they have accepted the treaty's provisions. The US, or any other country that does not ratify the agreement, can avoid the legal consequences of not reducing its GHG emissions but may not be able to avoid social condemnation by the international community – and perhaps its own polity – for failing to adopt climate-friendly policies. The condemnation of India and Pakistan for their nuclear policies (although neither has joined the Nuclear Nonproliferation Treaty), and of Norway for whaling that is legally compliant with the whaling treaty, illustrates how international norms can simultaneously stem from but not be constrained by the compliance definitions of international legal instruments. These pressures are reinforced by regular meetings of regime bodies (and associated media coverage), as well as by periodic reports from the Intergovernmental Panel on Climate Change (IPCC), other scientific bodies, NGOs, and statements by governments and international bodies about the need to avert climate change. 'Strictly economic decisions' become increasingly viewed as 'economic decisions that contribute to climate change'. 'Economic decisions that contribute to climate change impacts' are increasingly viewed as socially inappropriate.

These processes, if sufficiently widespread, are likely to induce a deep transformation in the values that people, and the states they populate, hold. They will reinforce the convictions of committed states that action to protect the climate is justified and warranted. They will increase the ranks of committed states by convincing at least some contingent states that compliance is the right thing to do even if it is costly, rather than simply when it is less costly than non-compliance. It will cause some resistant states to re-evaluate their assessment of costs and benefits within a context in which climate protection receives greater attention and praise than it does currently, leading them to join the regime during the second commitment period as contingent states. Such a transformation is by no means assured. However, the flexibility mechanisms incorporated into the climate regime help make such a transformation more likely. Whether it occurs will depend on whether the many other determinants of norms collectively foster or impede such a transformation.

## Conclusion

Will the nations of the world achieve the initial goals they set for themselves in the FCCC and, more importantly, the goals they will need to set to achieve a significant slowing in the rate of climate change? Many years will need to pass before any serious assessment can be made of that issue. Indeed, the nature of the climate change problem means that the regime will never be able to solve the climate change problem but will, at best, find ways to manage it over time (Clark, 1989). The initial obligations and compliance mechanisms established under the regime appear to have laid a useful foundation for progress in that direction. The enforcement mechanisms seem unlikely to be used frequently. Yet the flexibility granted to governments in when and how to meet their targets seems likely to induce high compliance with first commitment period targets. Some countries will make significant and costly efforts to meet those targets. More states are likely to take advantage of the regime's flexibility to comply without making significant domestic emission reductions, acquiring emission credits that are likely to be quite cheap because of the availability of hot air credits. In the short term, legal compliance with limited behavioural change is likely to produce only small changes in the global trajectory of GHG emissions.

This pessimism is mitigated, however, by two factors. First, some states will respond to the regime's requirements by adopting policies and behaviours that they would not have otherwise adopted. Some new policies to encourage conservation will be tried. Some more efficient technologies for producing and using energy will be developed. Some more climate-attentive approaches to land use, land-use change and forestry will be explored . Although these changes are unlikely to be significant before the end of the first commitment period, they nonetheless will contribute in small but immediate ways to averting climate change and provide the foundation for social learning that can contribute in much greater ways in the future (Social Learning Group 2001a; 2001b).

Second, and far more importantly, creating high compliance with the Kyoto targets (even if it is empty compliance without corresponding behaviour changes) will foster a shift in normative dialogue regarding behaviours that contribute to climate change. Such a shift to climate change behaviours being assessed within a logic of appropriateness is essential if governments and private actors are going to engage in the behaviours needed to avert climate change: behaviours that will, undoubtedly, be costly when measured in strictly material terms. The regime must – and its flexibility mechanisms make it more likely to – progressively convince a wide range of currently hesitant and resistant actors that acting to avert climate change is worthwhile even when the immediate costs of doing so are high. This dynamic social process of reframing climate protection as the only appropriate behaviour will help to establish a regime in which most actors focus on making concerted efforts to prevent climate change, regardless of whether those actions fall short of or go beyond some legal definition of compliance. Thus, the flexibility mechanisms that, at present, appear only to make empty compliance more likely may, over time, initiate social processes that lead to deep-seated normative changes that, in turn, may produce the dramatic, long-term changes in human behaviours that are necessary to avert climate change.

## Notes

1 This chapter is a significantly revised version of 'Institutional Aspects of Implementation, Compliance, and Effectiveness' in Luterbacher, U. and Sprinz, D. (eds 2001) *International*

*Relations and Global Climate Change*, Massachusetts Institute of Technology, Cambridge, MA. Reprinted by permission of the publisher. The current version has benefited greatly from comments from Thomas Gehring, Olav Schram Stokke, David Victor, the editors and contributors to the earlier volume, and contributors to the current volume. Much of my thinking on these issues has been influenced by Abram Chayes and Antonia Handler Chayes, Oran Young and Edward Parson. I wish to express my appreciation to all these scholars for their insights on these issues. The writing of this chapter was supported by a 2002 Summer Research Award from the University of Oregon and by a Sabbatical Fellowship in the Humanities and Social Sciences from the American Philosophical Society.

2 United Nations Framework Convention on Climate Change, Art 2.

3 Kyoto Protocol, Art 3.

4 Marrakesh Accords, Decision 24/CP.7, Annex, Sec XV

5 Id, Sec, XIII.

6 Marrakesh Accords, Decision 18/CP.7, Annex, para 6.

7 Whether the Marrakesh Accords permit unilateral CDM has been a contentious issue. The relevant text is found in Decision 17/CP.7, Sec G, para 40(a) : 'The designated operational entity shall: (a) Prior to the submission of the validation report to the executive board, have received from the project participants written approval of voluntary participation from the designated national authority of each Party involved, including confirmation by the host Party that the project activity assists it in achieving sustainable development.'

8 Marrakesh Accords, Decision 24/CP.7, Annex, Secs IV and XIV.

9 Id, Sec XV.

10 Marrakesh Accords, Decision 16/CP.7, Annex, Appendix B.

11 Id, Annex, Sec E.

12 Marrakesh Accords, Decision 24/CP.7, Annex, Sec XIII.

13 Id, Sec IV.

14 Id, Sec XIV.

## References

Axelrod, R. and Keohane, R. O. (1986) 'Achieving Cooperation Under Anarchy: Strategies and Institutions' in Oye, K. (ed) *Cooperation Under Anarchy*, Princeton University Press, Princeton, NJ, 226–54

Brown Weiss, E. and Jacobson, H. K. (eds 1998) *Engaging Countries: Strengthening Compliance With International Environmental Accords*, MIT Press, Cambridge, MA

Chayes, A. and Chayes, A. H. (1995) *The New Sovereignty: Compliance With International Regulatory Agreements*, Harvard University Press, Cambridge, MA

Chayes, A. H., Chayes, A. and Mitchell, R. B. (1995) 'Active Compliance Management in Environmental Treaties' in Lang, W. (ed) *Sustainable Development and International Law*, Graham and Trotman, London, 75–89

Clark, W. C. (1989), 'Managing Planet Earth', *Scientific American*, vol 261, 46–58

Downs, G. W., Rocke, D. M. and Barsoom, P. N. (1996) 'Is the Good News About Compliance Good News About Cooperation?', *International Organization*, vol 50, 379–406

Finnemore, M. and Sikkink, K. (1998) 'International Norm Dynamics and Political Change', *International Organization*, vol 52, 887–917

Keck, M. E. and Sikkink, K. (1998) *Activists Beyond Borders: Advocacy Networks in International Politics*, Cornell University Press, Ithaca, NY

Keohane, R. O. and Levy, M. A. (eds 1996), *Institutions for Environmental Aid: Pitfalls and Promise*, MIT Press, Cambridge, MA

King, G., Keohane, R. O. and Verba, S. (1994) *Designing Social Inquiry: Scientific Inference in Qualitative Research*, Princeton University Press, Princeton, NJ

Krasner, S. D. (1983) 'Structural Causes and Regime Consequences: Regimes as Intervening Variables' in Krasner, S. D. (ed) *International Regimes*, Cornell University Press, Ithaca, NY, 1–22

March, J. and Olsen, J. (1998) 'The Institutional Dynamics of International Political Orders', *International Organization*, vol 52, 943–70

Michaelowa, A. (2001) *Rio, Kyoto, Marrakesh: Groundrules for the Global Climate Policy Regime*, Hamburg Institute of International Economics, Hamburg

Morlot, J. C. (1998) 'Monitoring, Reporting, and Review of National Performance Under the Kyoto Protocol', OECD Information Paper, OECD, Paris

Pew Center on Global Climate Change (2001) *Summary of the Marrakesh Accords on Climate Change*, Pew Center on Global Climate Change, Washington, DC

Social Learning Group (ed 2001a) *Learning to Manage Global Environmental Risks, Volume 1: A Comparative History of Social Responses to Climate Change, Ozone Depletion and Acid Rain*, MIT Press, Cambridge, MA

Social Learning Group (ed 2001b) *Learning to Manage Global Environmental Risks, Volume 2: A Functional Analysis of Social Responses to Climate Change, Ozone Depletion and Acid Rain*, MIT Press, Cambridge, MA

Sprinz, D. and Vaahtoranta, T. (1994) 'The Interest-based Explanation of International Environmental Policy', *International Organization*, vol 48, 77–105

Victor, D. G. (2001a) 'Commentary on the "new Pronk" text', unpublished manuscript, Council on Foreign Relations, New York

Victor, D. G. (2001b) 'Commentary on the negotiations at the Hague and the Pronk synthesis text', unpublished manuscript, Council on Foreign Relations, New York

Victor, D. G., Nakicenovic, N. and Victor, N. (1998a) *The Kyoto Protocol Carbon Bubble: Implications for Russia, Ukraine, and Emission Trading (IR-98-094)*, International Institute for Applied Systems Analysis, Laxenburg, Austria

Victor, D. G., Raustiala, K. and Skolnikoff, E. B. (eds 1998b) *The Implementation and Effectiveness of International Environmental Commitments*, MIT Press, Cambridge, MA

Victor, D. G. and Salt, J. E. (1994) 'From Rio to Berlin: Managing Climate Change', *Environment*, vol 36, 6–15, 25–32

Young, O. R. (1992) 'The Effectiveness of International Institutions: Hard Cases and Critical Variables' in Rosenau, J. N. and Czempiel, E.-O. (eds) *Governance Without Government: Change and Order in World Politics*, Cambridge University Press, New York, 160–94

Chapter 4

# Reporting and Verification of Emissions and Removals of Greenhouse Gases

Terje Berntsen, Jan Fuglestvedt and Frode Stordal

## Introduction

For a compliance regime to be effective, it must be possible to verify that reductions actually have taken place according to the commitments undertaken. In the political negotiations of the Framework Convention on Climate Change (FCCC) there was a demand for measures to be both comprehensive and cost effective (UNFCCC, Art 3.3). These dual demands were made operational in the 1997 Kyoto Protocol by formulating emissions targets as the aggregate anthropogenic carbon dioxide ($CO_2$) equivalent emissions of six specified greenhouse gases or groups of gases: $CO_2$, methane ($CH_4$), nitrous oxide ($N_2O$), hydrofluorocarbons (HFCs), perfluorocarbons (PFCs) and sulphur hexafluoride ($SF_6$).[1] However, the inclusion of several gases from a variety of sources, including managed ecosystems and even carbon sinks in forests, increases the uncertainty in emission estimates considerably. Thus there is a trade-off between a cost effective and comprehensive agreement, and the possibility to assess compliance with obligations.

The Kyoto Protocol requires that emission inventories used to assess compliance with Protocol commitments shall be based on the best available scientific knowledge and be prepared according to the detailed guidelines, good practice guidance and reporting instructions for national inventories of greenhouse gas (GHG) emissions developed by the Intergovernmental Panel on Climate Change (IPCC) (IPCC, 1996b; IPCC, 2000). These include guidance for handling uncertainties, quality assurance/quality control and options for the verification of national emission data using independent methods. The control and verification of national inventories is performed by an international Expert Review Team (ERT),[2] which assesses the quality of each inventory. The reviewed inventory of estimates of emissions and removals will form the basis for assessing compliance in legal terms. The fact that emission estimates form the basis for assessing compliance is an important point, since it implies that under the Kyoto Protocol it is possible for a party to be in legal compliance while not in scientific compliance. Real emissions are not fully known, and in some cases there is a danger that the reported emissions and removals can be very different from the reality. This means that the estimated and reported emissions/removals can show fulfilment of the commitments, while actual emissions might exceed the agreed target. If this is due to shortcomings in the guidelines (for example, unknown sources and large uncertainties in emission factors) used to prepare national inventories as defined under Article 7 of the Kyoto Protocol, the party would still be in legal compliance, even if the actual emissions exeed the party's commitments. The most dangerous place for this gap to occur is in the area of emissions/sinks from managed ecosystems (eg $N_2O$ emissions from agriculture), where the emissions/sinks can be highly variable (geographically and temporally) and are largely dependent on factors such as fluctuations in the local climate. The consequence can be that the IPCC guidelines propose biased emissions factors. In the worst case, there might be unknown emission sources.

In addition to assessing compliance from national GHG inventories, there are prospects for developing a methods commonly called inverse modelling (IM) based on accurate measurements of atmospheric concentrations and numerical transport modelling, which in principle can be used to calculate the origin of net emissions of trace gases. Since these methods calculate actual emissions, they can only be used to assess scientific compliance, or the effectiveness of the regime (see also Chapter 3 by Mitchell). The advantage with IM is that it is completely independent of subjective judgements about sources, which form an unavoidable element of the preparation of each party's inventory. At the current stage of development, however, there are several obstacles that need to be resolved before the uncertainties and geographical resolution in the calculated distribution of emissions reach a level that is acceptable for use in a compliance regime. These uncertainties originate from insufficient measurement networks, inaccurate measurement techniques, lack of knowledge of natural sources and sinks, insufficient quality of meteorological data, and coarse representations of key transport processes in transport models. This chapter will discuss the principles of various methods and approaches for verification of GHG emissions, how the uncertainties can be reduced, and – most importantly – how the remaining uncertainty can be dealt with within a compliance regime.

The chapter is organized as follows. The initial section provides an overview of the main properties of the GHGs included in the Kyoto Protocol, and how they affect the radiative forcing of the climate system. Next, the two main approaches for assessing compliance with the Kyoto Protocol are discussed: self-reporting through national inventories, which forms the basis for assessing the legal compliance with the agreement; and possibilities for using independent methods based on IM. Finally, the last section discusses how uncertainties affect these two methods, and proposes a possible way to combine the two methods in a compliance regime.

## Anthropogenic interference with the climate system

Since pre-industrial times, the atmospheric concentrations of several important GHGs have increased dramatically. The global average concentration of $CO_2$ has increased from 280 parts per million by volume (ppmv) to a present level of 365 ppmv (+30 per cent). The level of $CH_4$ has risen from 700 parts per billion by volume (ppbv) to 1745ppbv (+150 per cent). $N_2O$ concentrations have increased by 17 per cent. These gases have large natural sources, but the increases in atmospheric concentrations are mainly due to anthropogenic activities. The dominant source of man-made emissions of $CO_2$ is fossil fuel combustion, while deforestation and changes in land use have also contributed significantly. The anthropogenic sources are small compared to the natural fluxes (less than 5 per cent), but it is well established that it is these man-made emissions that have caused the accumulation in the atmosphere, since natural emissions are balanced by natural sinks. Important man-made sources of $CH_4$ include ruminants, rice agriculture, burning of biomass, landfills, and production and consumption of fossil fuels. Agriculture and industrial processes are important anthropogenic sources of $N_2O$. In addition, man-made activities also result in emissions of gases that were not present in the pre-industrial atmosphere. Table 4.1 shows the changes in concentrations for some important GHGs.

**Table 4.1 Changes in the concentrations of some important GHGs**

| | $CO_2$ | $CH_4$ | $N_2O$ | $SF_6$ | HFC-23 | $CF_4$ |
|---|---|---|---|---|---|---|
| Pre-industrial concentration | Approx 280ppmv | Approx 700ppbv | Approx 270ppbv | 0 | 0 | 40pptv |
| Present concentration | 365ppmv | 1745ppbv | 314ppbv | 4.2pptv | 14pptv | 80pptv |
| Rate of concentration change | 1.5ppmv/yr | 7ppbv/yr | 0.8ppbv/yr | 0.24pptv/yr | 0.55pptv/yr | 1pptv/yr |
| Lifetime | 5–200 years* | 12 years | 114 years | 3–200 years | 260 years | >50,000 years |

*Source*: IPCC, 2001

*Notes*: HFC-23 = $CHF_3$; $CF_4$ = carbon tetrafluoride; ppmv (parts per million by volume)= number of molecules of a gas per 1 million molecules of air; ppbv (parts per billion by volume) = number of molecules of a gas per 1 billion ($10^9$) molecules of air; pptv (parts per trillion by volume) = number of molecules of a gas per 1 trillion ($10^{12}$) molecules of air.

* No single lifetime can be given for this gas since its removal is controlled by several different processes with different time constants.

The timescale for the mixing of gases within a hemisphere is a few months, while mixing between the hemispheres has a timescale of two years. Gases with atmospheric lifetimes longer than this will thus be well mixed in the troposphere.[3] All the gases shown in Table 4.1 will thus be homogeneously mixed in the troposphere. Geographical gradients in concentrations will generally be quite small, which makes it difficult to detect emissions from a given country with great certainty. For gases with shorter lifetimes than the tropospheric mixing times, there will be regional variations in concentrations depending on the location of the sources. This is the case for nitrogen NOx and carbon monoxide (CO), both important precursors of tropospheric ozone ($O_3$). With a lifetime in the order of 100–200 days, tropospheric ozone itself also shows significant regional variations. It is also the case for sulphate particles and their precursor sulphur dioxide ($SO_2$), which act as cooling agents and have lifetimes of only a few days, resulting in large regional variations in atmospheric levels.

The observed changes in concentrations of the gases mentioned above have caused 'radiative forcing'[4] of the climate system. Radiative forcing is a metric that can be used to assess the strength of a climate change mechanism, and is commonly used for comparing the impact on the climate of various gases and substances. A positive forcing indicates warming, while a negative forcing means cooling. Figure 4.1 shows the present radiative forcing due to changes since pre-industrial times for a number of forcing agents.

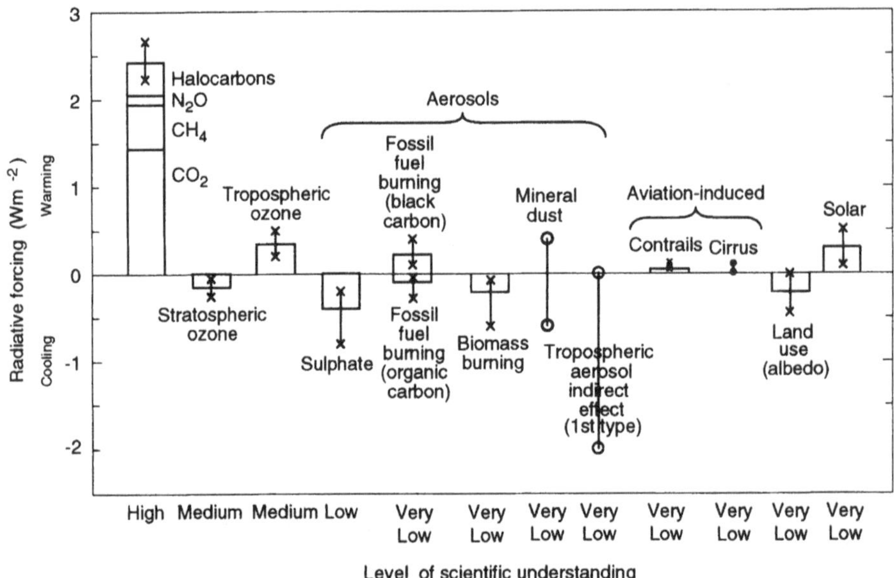


**Figure 4.1 The global mean radiative forcing of the climate system from various agents for the year 2000 relative to 1750**

*Source*: IPCC, 2001

*Note*: albedo means the fraction of incoming solar radiation reflected out to space

The largest radiative forcing is caused by $CO_2$,while $CH_4$, $N_2O$ and halocarbons (mainly CFCs and other gases controlled by the Montreal Protocol) also contribute significantly. In addition to the forcing from these well-mixed GHGs, there are also significant effects of other gases and substances. Reductions in stratospheric[5] ozone ('the ozone layer') have caused a negative forcing, while increases in troposphere ozone have led to a warming effect. Aerosols (airborne particles) have in most cases caused cooling due to their scattering of solar radiation, and through effects on clouds. Under the current climate regime (the Kyoto Protocol), only emissions contributing to the leftmost bar are regulated.

This illustration of man-made radiative forcing of the climate system shows us several things:

1 Of the man-made effects on climate, increased levels of $CO_2$ are the most important.
2 There are several other forcing agents in addition to the well-mixed gases with both cooling and warming effects.
3 The level of scientific understanding regarding the forcing from the well-mixed gases is considered high, while for many potentially important forcing agents the understanding is considered to be limited.

The figure also indicates that the well-mixed gases together cause the largest warming effect, and halocarbons are responsible for a large part of this. Many of these gases are controlled by the Montreal Protocol. Table 4.2 shows the radiative forcing since pre-industrial times for the gases controlled by the Kyoto Protocol.

**Table 4.2 Importance of the various Kyoto gases in terms of radiative forcing since pre-industrial times, and relative importance of current emissions in terms of $CO_2$-equivalents**

| Gas | Radiative forcing since pre-indiustrial times ($W/m^2$) | Contribution since pre-industrial times | Relative importance of current emissions (global) | Relative importance of current emissions (within OECD) |
|---|---|---|---|---|
| $CO_2$ | 1.46 | 69.56% | 67.5% | 72.6% |
| $CH_4$ | 0.48 | 22.87% | 15.7% | 9.6% |
| $N_2O$ | 0.15 | 7.15% | 15.7% | 15.7% |
| $CF_4$ | 0.003 | 0.14% | 0.21% | 0.29% |
| $C_2F_6$ | 0.001 | 0.05% | | |
| $SF_6$ | 0.002 | 0.10% | 0.35% | 0.52% |
| HFC-23 | 0.002 | 0.10% | 0.58% | 1.3% |
| HFC-134a | 0.001 | 0.05% | | |
| HFC-152a | 0.000 | 0% | | |
| TOTAL | 2.099 | 100% | 100% | 100% |

*Source*: IPCC, 2001

*Notes*: HFC-23 = CHF3; HFC-134a = CH2FCF3; HFC-152a = CH3CHF2

The radiative forcing of the climate system has probably already caused global warming and much of the observed increase in temperature since 1860 can, according to IPCC, 2001, be attributed to increased levels of $CO_2$ and the other GHGs. This warming has been counteracted by increased levels of particles that have caused negative radiative forcing and a cooling effect (see Figure 4.1).

In the future, the relative importance of the well-mixed gases is likely to increase. Due to their long lifetimes, there are generally significant imbalances between their sources and sinks. This means that even if the emissions were kept constant at today's level, the concentrations of these gases would still increase for several decades (IPCC, 2001). Also, due to local pollution and health issues, emissions resulting in aerosols and tropospheric ozone are likely to be controlled.

Scenario studies given in IPCC, 2001 show that during this century the global mean temperature can be expected to increase by 1.4 to 5.8°C, which will bring the global mean temperature far above the range observed over the last 1000 years.

## The 'Kyoto basket'

In the Kyoto Protocol, emissions targets are defined as the 'aggregate anthropogenic carbon dioxide equivalent emissions' of six specified GHGs (Art 3.1), which means that the emissions of a set of GHGs with very different atmospheric lifetimes and radiative properties are converted to interchangeable units on one common scale. The Kyoto gases are $CO_2$, $CH_4$, $N_2O$, HFCs, PFCs and $SF_6$. The Protocol gives a specific emissions target relative to 1990 emission levels for each party included in Annex B of the Protocol. The aggregate emissions target is 5.2 per cent below 1990 emission levels and is to be met during the period 2008–12. The flexible approach of the Protocol allows each party to determine which gases it will focus on in its abatement strategy. Under Article 5 of the Kyoto Protocol, the global warming potentials[6] (GWPs) given in IPCC (1996a) for a time horizon of 100 years will be used to convert the gases in the Protocol to the common unit of $CO_2$ equivalents. $GWP_{100}$ values and atmospheric lifetimes are given for some selected gases in Table 4.3.

**Table 4.3 GWP100 values from IPCC (1996a) for the GHGs included in the Kyoto Protocol**

| Gas | Lifetime (years) | $GWP_{100}$ |
|---|---|---|
| HFC-152a | 1.5 | 140 |
| HFC-32 | 5.6 | 650 |
| HFC-134a | 14.6 | 1,300 |
| $CH_4$* | 12.2 | 21 |
| HFC-125 | 32.6 | 2,800 |
| $CO_2$** | Variable | 1 |
| $N_2O$ | 120 | 310 |
| $SF_6$ | 3200 | 23,900 |
| $C_2F_6$ | 10,000 | 9,200 |
| $CF_4$ | 50,000 | 6,500 |

*Notes*: HFC-32 = $CH_2F_2$; HFC-125 = $CHF_2CF_3$

* The GWP for $CH_4$ includes indirect effects on tropospheric $O_3$ and stratospheric water.

** Derived from the Bern carbon cycle model.

When the emissions of the various gases are multiplied by their respective GWPs, the emissions are converted to a common unit, often called $CO_2$-equivalents. The common definition of the $CO_2$-equivalent is:

$$CO_2\text{-eq}(H) = GWP_i(H) \cdot E_i$$

where $GWP(H)_i$ is the global warming potential of gas i; $E_i$ represents the emission of gas i measured by mass; and $CO_2$-eq(H) is the $CO_2$-equivalent amount of gas i using GWPs for a time horizon H.

## Two main approaches

A system for assessing compliance with the obligations for anthropogenic emissions and sinks for a party under the climate convention may be based on two main approaches:

1 National emissions inventories (including sinks) reported from each party, with a subsequent independent expert review.
2 IM techniques based on numerical transport models, observations of concentrations in the atmosphere and inventories of emissions from natural sources.

With respect to the first approach, which will be the basis for assessing compliance by each party in the first commitment period (under Article 7 of the Kyoto Protocol), the FCCC calls on the parties to:

1 Develop, update periodically, publish and make available to the Conference of the Parties (COP) their national inventories of anthropogenic emissions by sources and removals by sinks of all GHGs not controlled by the Montreal Protocol.
2 Use comparable methodologies for inventories of GHG emissions and removals, to be agreed upon by COP.

The IPCC Guidelines were first accepted in 1994 and published in 1995. UNFCCC COP-3, held in 1997 in Kyoto, reaffirmed that the *IPCC Guidelines for National Greenhouse Gas Inventories* (IPCC, 1996b) should be used as 'methodologies for estimating anthropogenic emissions by sources and removals by sinks of greenhouse gases' in calculation of legally binding targets during the first commitment period. In addition to these guidelines, the IPCC has also prepared a report entitled *Good Practice Guidance and Uncertainty Management in National Greenhouse Gas Inventories* (IPCC, 2000)[7] at the request of the Subsidiary Body for Scientific and Technological Advice at its eighth session (SBSTA-8). This report was approved by COP-6.

Although this self-reporting approach provides a common set of methodologies and handling of uncertainties and verification by an independent Expert Review Team,[8] reporting emission inventories at a national level creates a possibility of systematic underreporting of emissions in the commitment period. In other words, parties may cheat by underreporting emissions to appear as if they are in compliance with the Protocol. The independent control provided by the Expert Review Teams can reduce, but not totally eliminate, the risk of systematic underreporting emissions. This is why the second approach has been suggested. Instead of being based on self-reporting, compliance could be assessed from a verification system independent of national reports. Such verification could be inversion techniques using data from a network of measurements of atmospheric concentrations of GHGs and modelling of transport and deposition of GHGs. The main output from inverse models is net emission estimates (sources minus sinks) on a geographical grid. The grid cells will in general be squares that do not follow the borders between nations. To be applicable, the grid cells must be small enough so that so that grid cells overlapping two parties do not make a significant fraction of the area, i.e. the grid cells can be summed up to represent a party.

The main advantage of this second approach is that it does not involve any reporting on a national level (except for that portion of national commitments that are obtained through the Kyoto mechanisms of emissions trading, joint implementation (JI) and the clean development mechanism (CDM)). The opportunity to cheat by systematic underreporting is thus removed. One disadvantage of this approach, however, is that due to the current uncertainties associated with IM, the possibility of legal action against a party based solely on modelling results is unlikely in the foreseeable future. At the same time, uncertainties embedded in the IM technique might make it possible for a party to choose to not fulfil its obligations and hope that this failure will not be detected with a reasonable level of uncertainty (cf the discussion section at the end of the chapter). Even for the purely man-made gases regulated under the Montreal Protocol (CFCs and HCFCs), assessing compliance is currently based on national inventories and not on IM.

With the current definition of compliance in the Kyoto Protocol, the assessment of 'scientific compliance' provided by IM cannot be applied directly. However, as will be discussed in the final section, we believe that there are possibilities to utilize the information from IM, in combination with national inventories and expert review, to enhance the quality of national inventories and to enhance the chances of detection of non-compliance.

The next two sections discuss the self-reporting and inversion approaches in more detail.

## Reporting of emission inventories

The national emission inventories are developed on the basis of the guidelines provided by the IPCC (IPCC, 2000). The quality of the reported emissions strongly depends on the quality of the underlying data. The *Good Practice Guidance* report from the IPCC suggests standard methods at different 'levels of sophistication' (Tiers, cf Figure 4.2) for estimating total emissions. The choice of method shall take into account the amount and quality of data available as well as the relative importance of each emission source. This includes choice of methodology, emission factors, activity data, uncertainties, quality assurance and quality control procedures, and time series consistency. Figure 4.2 shows an example of a decision tree for selecting the method for estimating $CO_2$ emissions from stationary combustion.

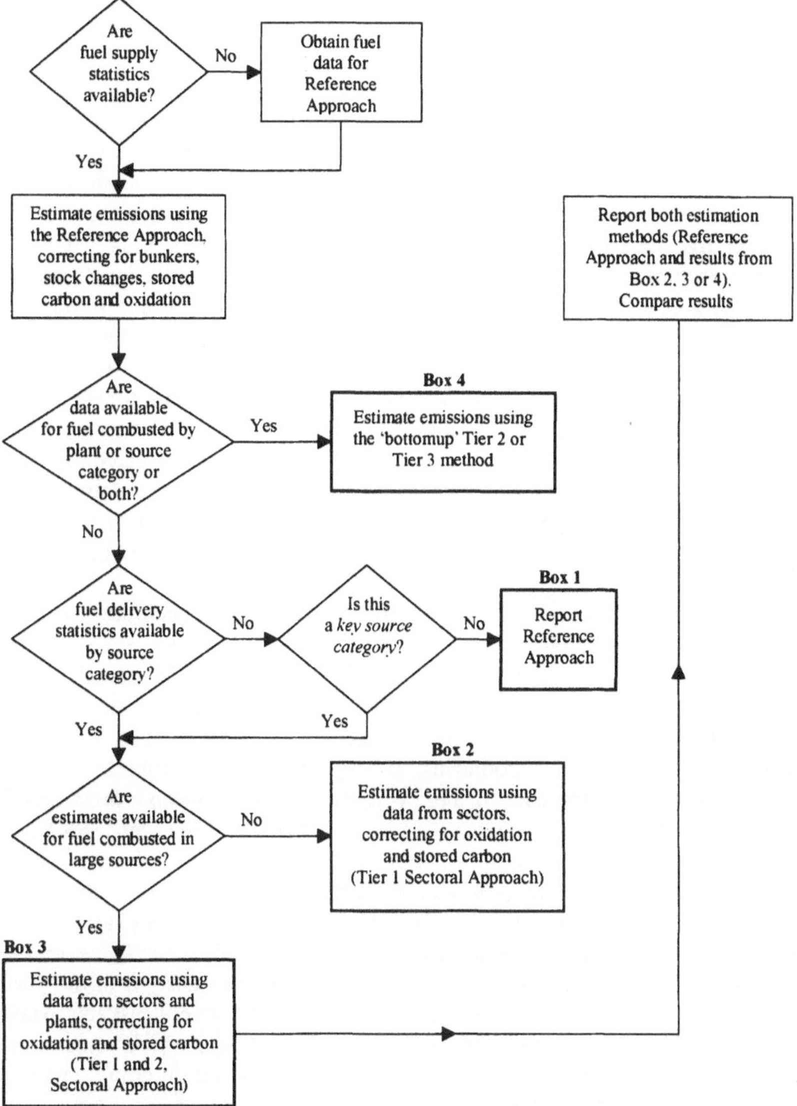


**Figure 4.2 Example of decision tree for selecting the method for estimation of CO2 emissions from stationary combustion**

*Notes*: Tiers are alternative methods for estimating emissions. Tier 3 denotes the most complex and accurate method.

* A key source category is one that is prioritized within the national inventory system because its estimate has a significant influence on a country's total inventory of direct GHGs in terms of the absolute emissions, the trend in emissions or both.

Similar decision trees are given for all relevant anthropogenic emissions sources for the Kyoto gases. For ozone precursors (CO, NOx and volatile organic compounds (VOCs)) this is given in the IPCC guidelines (IPCC, 1996b). It should be noted that the methodology includes an assessment of the relative importance of each source category to assure that limited resources are prioritized to the most relevant emissions. Results from IM could be used to identify important sources.

Although the guidelines provide standard methods for uncertainty analysis and quality assurance/quality control, it is clear that these uncertainties will vary considerably between sources, gases and nations. The reason is that some emissions sources are complex and can be highly variable in space and time, and there is a lack of representative measurements and models (see also Chapter 3 by Mitchell). It is, however, important to notice that uncertainties in the absolute levels of emissions differ, and generally will be larger than the uncertainties regarding trends over time. Since the Kyoto commitments are given in relative (percentage) terms compared to the base year (1990), it is the latter uncertainty that is most important for assessing compliance. Generally, the estimates will be more uncertain for emissions from sources involving biological systems (e.g. $CO_2$, $N_2O$ and $CH_4$ emissions and sinks from agriculture and forestry). An overview of the range of estimated uncertainties is given in Table 4.4. During the first commitment period, uncertainties in the 1990 emission inventories will be difficult to reduce, but for later periods improved consistency in the methodology and quality of the data over time should reduce the uncertainties.

To assure that the reported inventories are complete, accurate and conform to the guidelines, independent Expert Review Teams (ERTs) will review the national inventories (see Chapter 2 by Ulfstein and Werksman and Chapter 3 by Mitchell). The ERTs will consist of four to five persons coordinated by the Secretariat and selected from a roster of experts nominated by the parties. They will also include experts from non-Annex I countries. If any problems are found, the ERTs may recommend adjusting the data. If there is disagreement between a party and the ERT about the data adjustment, the Compliance Committee will intervene. ERTs also have the mandate to raise any apparent implementation problems – known as 'questions of implementation' – with the Compliance Committee. To facilitate improvements in the review process and to ensure that reviews are performed consistently and in time for all countries, a standing group of ERTs has been established. The extent to which the ERTs will be able to detect non-compliance is a key question in the verification procedure of the Kyoto Protocol. This will depend on several aspects, such as the nature of the measures adopted by the party (which gases, which source categories, etc.). For some sectors, reported emissions might be based on confidential information which could be difficult to verify.

Rypdal and Winiwarter (2001) have reviewed estimates of uncertainties in current GHG inventories (total emissions and trends, not including forest emissions/sinks of $CO_2$) for high quality inventories from five industrialized countries (Austria, the UK, the US, Norway and the Netherlands).[9] Uncertainties in the total emissions are found to be 5–20 per cent (see Table 4.4), and the largest contribution to the uncertainty comes from subjective estimates of the uncertainty of $N_2O$ emissions from agricultural soils. The estimated uncertainties in the trends are lower (4–5 percentage points) than in the totals, since some of the systematic errors are likely to affect the estimates throughout the period. This means that refined methods that can reduce the systematic errors will not have significant effects on the uncertainties of the trend estimates. Random uncertainties in trends may be as high as 3 percentage points (Winiwarter and Rypdal, 2002).

**Table 4.4 Assessed uncertainties in emissions of each Kyoto GHG, total emissions and trend (percentage points)**

| | Austria | Norway | The Netherlands | UK | USA | IPCC (1996) |
|---|---|---|---|---|---|---|
| $CO_2$ | 2 | 3 | 3 | 4 | 3 | 10 |
| $CH_4$ | 48 | 22 | 17 | 17 | 36 | 30 |
| $N_2O$ | 90 | 200 | 34 | 230 | 120 | |
| HFCs | – | 50 | 41 | 24 | 25 | |
| PFCs | – | 40 | 100 | 20 | – | |
| $SF_6$ | – | 5 | 50 | 13 | – | |
| Total | 10 | 21 | 4.4 | 19 | 13 | |
| Total trend (1990–2010) | 5 | 4 | – | 4 | – | |

*Source*: Rypdal and Winiwarter, 2001

*Note*: Totals refer to combined uncertainties of GHG weighted with their GWP100 values. Numbers are two standard deviations (95 per cent) as a percentage of total emissions. Trend in percentage points as two standard deviations. (See Rypdal and Winiwarter, 2001, for details and references). Emissions from LULUCF (land use, land-use change and forestry) are not included.

The uncertainties in the trends are about equal to typical obligations of the parties in the Kyoto Protocol. Thus, in cases where obligations are met with a margin less than the uncertainty, compliance may be questioned from a scientific point of view. It is very doubtful that parties will reduce their emissions by more than the commitments according to their best estimate. Therefore it does not seem probable that compliance with the Kyoto Protocol can be easily proven. This difficult problem will have to be addressed by the ERTs, which are to 'prepare a report ... assessing the implementation of the commitments of the party and identifying any potential problem in, and factors influencing, the fulfilment of commitments'.[10] For future commitment periods, with larger obligations and/or refined methods for estimating emission trends, this problem will probably be less severe.

## IM techniques

The well-mixed nature of GHGs has been an advantage for the negotiation of the Protocol, because the effects of emissions are independent of the location of the emissions, thus allowing the same weights (GWPs) to apply for all participants (in contrast to the effects of short-lived species, eg $NO_x$ (a precursor of ozone) or $SO_2$ (a precursor to sulphate particles)). However, from the point of view of verification of emissions on a country-by-country scale, the well-mixed nature is a disadvantage because the small concentration gradients make it more difficult to assess the sources of the emissions.

The principle of an inversion model is as follows. From the knowledge of the spatial and temporal distribution of a GHG in the atmosphere (from observations), the circulation in the atmosphere (from numerical weather forecast models), and natural sources and sinks of the GHG, the location (in space and time) of anthropogenic emissions can be calculated (e.g. Bousquet et al, 1999; Kaminski et al, 1999; Houweling et al, 1999).

In mathematical terminology, a chemical transport model (CTM) defines a mapping (or function) of a parameter set (including emissions) on the simulated concentrations (Houweling et al, 1999). For any measured concentration $d_{i,t}$ at location *i* and time *t*, this function can be written in the form:

$$d_{i,t} = d_{i,t}[\mathbf{F},\mathbf{Q},\mathbf{C_0}]$$

where **F** is a vector which components $f_{j,s}$ are the net surface flux (emissions (natural + anthropogenic) – sinks) at location j at time s. **Q** is the loss in the atmosphere (zero for $CO_2$, but non-zero for other gases), and $\mathbf{C_0}$ is the initial concentration at the time of the simulation. The aim of the inversion is to find the set of the most probable values for the parameters $\mathbf{F},\mathbf{Q},\mathbf{C_0}$. The technique usually involves an *a priori* estimate of the parameters $\mathbf{F},\mathbf{Q},\mathbf{C_0}$ and the use of a global set of observed concentrations. The *a priori* estimates are not absolutely necessary, but it speeds up the calculations and thus makes it computationally possible to perform more detailed simulations. For emissions of GHGs, the *a priori* estimate will be emissions reported by parties to the convention.

In its most simple version, IM can be applied to global burdens, treating the atmosphere as a single box and neglecting the transport effects of GHGs. Neglecting transport means that uncertainties related to transport do not influence the results, thus yielding results with smaller uncertainties. Although the results from this kind of model give only global total emissions, they can be used as a check to see if the emissions reported by the parties are in agreement with total global emissions; in other words, they can be used to assess the effectiveness of the climate regime as a whole. As part of a system for verification of emissions, the model gives useful information about whether the total budget is correct, and thus can be used to determine whether further control of national inventories or improvement in the guidelines for reporting emissions are necessary.

Figures 4.3 and 4.4 give some examples of global budget calculations (Höhne and Harnisch, 2002) for HFC-134a (a CFC replacement gas) and for $SF_6$. The shaded regions in the figures show the reported emissions from various main emitting countries. The lines give the calculated global emissions based on atmospheric observations of concentrations and estimated loss. In the case of HFC-134a, there seems to be a reasonable agreement between the reported emissions, while for $SF_6$ there is obviously significant underreporting or a large unknown source.

### *Inversion on a regional and country-specific scale*

At a regional level, IM is carried out with models that either follow specific air parcels as they move along a given (calculated) trajectory and exchange tracers with the surface (Lagrangian models), or with Eulerian models, which divide the atmosphere into fixed grid boxes and then calculate the fluxes between the grid boxes as well as surface exchanges.

Inversion techniques based on global Eulerian models have been used successfully to study regional net emissions of $CO_2$ when the regions are of continental size. This has proved very helpful in constraining the total budget of $CO_2$. Gurney et al (2002) report estimated regional (22 regions globally) natural emissions/sinks of $CO_2$ based on the results from 16 CTMs. For $CO_2$, fossil fuel sources are currently much better known than natural sources and sinks, so the fossil fuel sources were specified in the calculations.

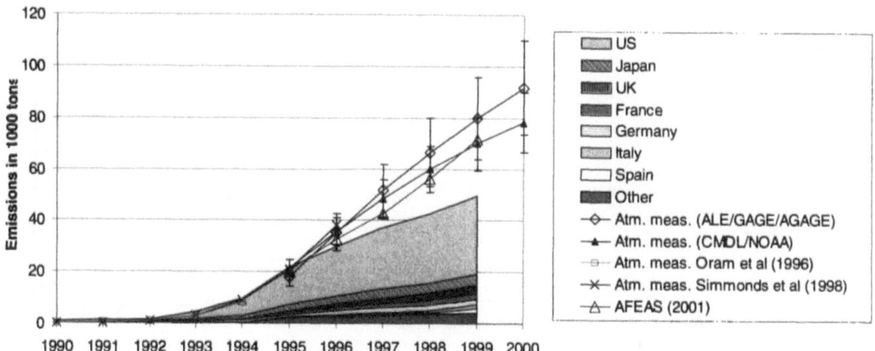


**Figure 4.3 Emission estimates of HFC-134a**

*Source*: Höhne and Harnisch, 2002

*Note*: See Höhne and Harnisch (2002) for explanation of the measurement programs ALE/GAGE/AGAGE, CMDL, NOAA and AFEAS

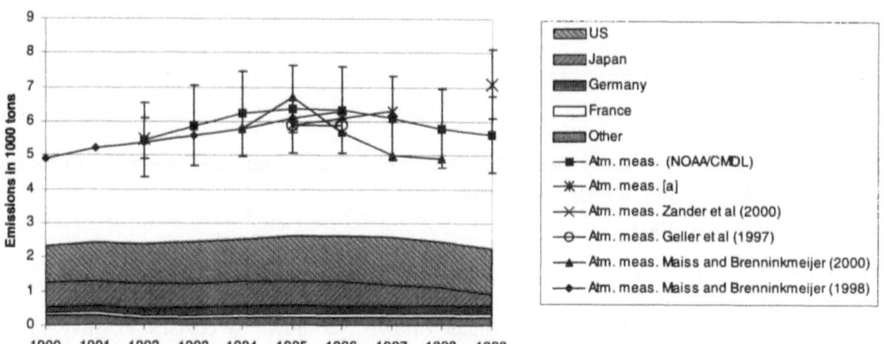


**Figure 4.4 Emission estimates of $SF_6$**

*Source*: Höhne and Harnisch, 2002

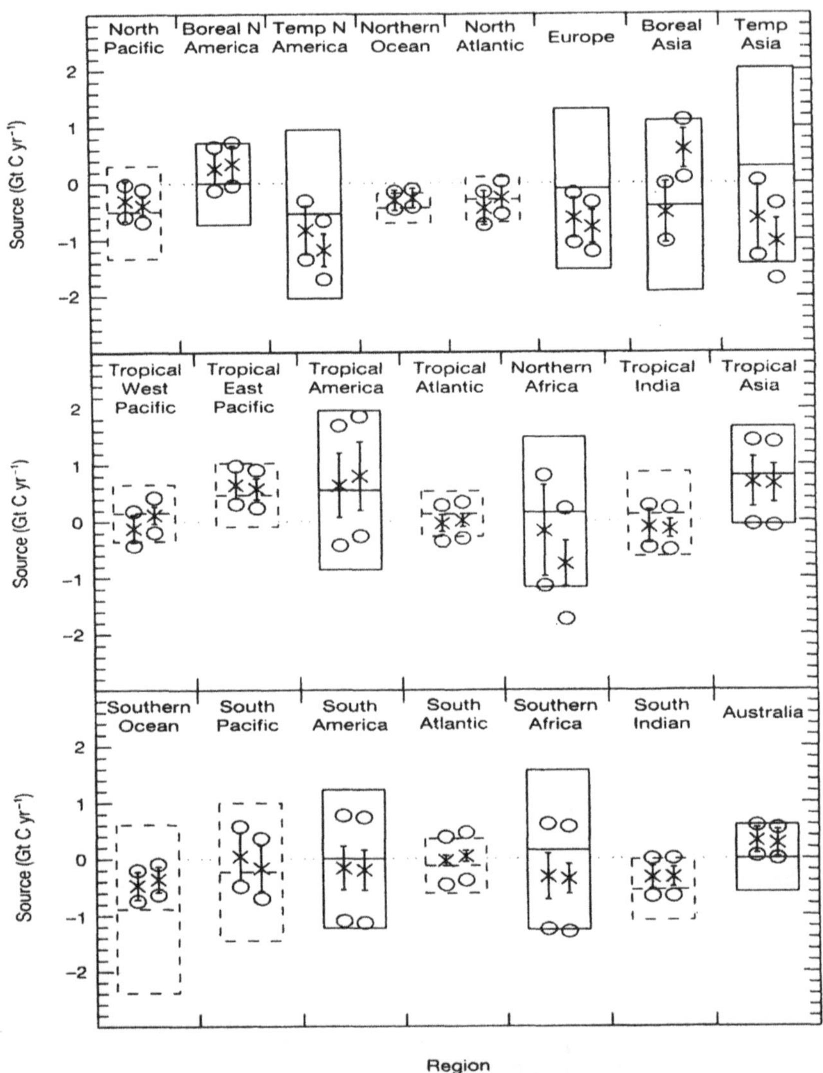


**Figure 4.5 Mean estimated net sources (except fossil fuel) and uncertainties for 22 regions across the globe**

*Source*: Gurney et al, 2002

*Notes*: Gt(C)/yr = $10^{12}$ kg carbon per year emitted; Temp = temperate; Based on two inversions with 16 CTMs. Left-hand symbols in each box are for an inversion including seasonal variation in the biosphere flux. Mean estimated fluxes are shown by crosses. Positive values indicate a source to the atmosphere. The prior flux estimates and their uncertainties are indicated by the boxes. The mean estimated uncertainty across all models (the 'within-mode' uncertainty) is indicated by the circles. The standard deviation of the models' estimated fluxes (the 'between-model' uncertainty) is indicated by the 'error bars'. See Gurney et al, 2002 for details about the calculations.

Figure 4.5 shows the results for the 22 regions. It should be noted that for the industrialized regions like temperate Northern America and Europe, the magnitude of the net fluxes and the uncertainties (both within model and inter model) are about equal to the fossil fuel emissions (≈1 Gt C $yr^{-1}$). To make the uncertainties comparable to the overall Kyoto commitments (-5.2 per cent), a decrease in uncertainty of a factor of 20 is needed. To reduce some of the uncertainties in the derived emission estimates, isotope techniques can be used. For $CO_2$ or $CH_4$, the fractions containing the carbon isotopes $^{13}C$ or $^{14}C$ can be measured, and separate inversions can be done for $^{12}CO_2$, $^{13}CO_2$ and $^{14}CO_2$ (or $^{12}CH_4$ $^{13}CH_4$, and $^{14}CH_4$ ) to constrain the estimated emissions. This is particularly useful since the various sources have different isotopic fingerprints, e.g. a very low fraction of $^{14}CO_2$ in fossil fuels due to radioactive decay.

To serve as a tool for verification of emissions from parties under the FCCC, the spatial scale of IM must be reduced so that emissions from separate parties can be identified, which implies that vertical mixing in atmosphere at timescales comparable to the timescales for horizontal transport from one country to another (hours to a few days) becomes more important. Representing this vertical mixing in the planetary boundary layer (PBL, the lower 50–2000m of the atmosphere) properly, and thus reducing the uncertainty in the estimated emissions to an acceptable level, requires significant improvements in the models used today. This is an area of active research, to bridge this gap between the local-scale flux measurements and continental-scale IMs.[11]

The EU-funded SOGE project (System for Observation of halogenatial Greenhouse gases in Europe)[12] has been initiated with the main objective of developing a new, cost-effective, long-term European observation system for halocarbons. The system will contribute to global observing networks. The results will support the Kyoto and the Montreal protocols in assessing the compliance of European regions with the Protocol requirements. In particular, the objectives of the observation system will be to:

- detect trends in the concentrations of greenhouse-active and ozone-destroying halocarbons;
- verify reported emissions and validate emission inventories for a series of halocarbons for Europe as a whole, as well as for certain regions; and
- develop observational capacity for all halocarbons included in the Kyoto Protocol (HFCs, PFCs and $SF_6$) for which this does not yet exist.

Figure 4.6 gives an example of a comparison between observed and modelled concentrations of CFC-12 at Mace Head (Ireland) using an Eulerian transport model. Based on the reported emissions, the air parcels pick up CFC-11 as they move over the source regions. Direct transport (with small dilution) from the source regions leads to enhanced concentrations. If the modelled peaks (red curve) are systematically lower than the observations during episodes with transport from a certain region, the results of the IM will be a quantification of the missing source in this region.

As discussed above, the outcome of an inverse model is a set of the most probable values for $F,Q,C_0$. For use in a compliance regime, it is the anthropogenic contribution to $F$ ($F_a$) that is of interest. For gases with large, variable and not very well-known natural sources and sinks (eg $CO_2$), the uncertainty in the estimate of the anthropogenic signal is enhanced. The natural sources and sinks ($F_n$) must either be specified through input data, or explicitly modelled as part of the CTM. Based on the knowledge of the uncertainty range in the other input parameters, an uncertainty range of $F_a$ can be estimated. In addition, the uncertainty connected to the mapping function (the CTM) must be included.

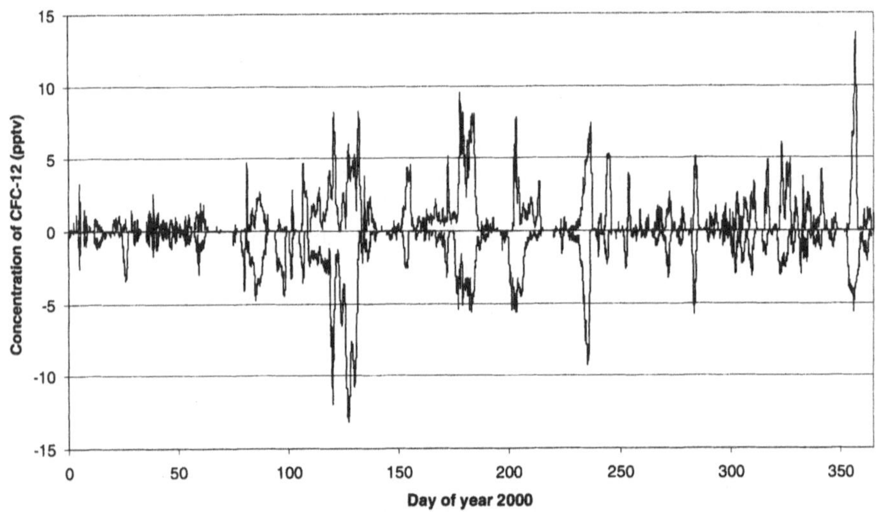


**Figure 4.6 Model predicted and observed concentrations (deviations from the baseline) of CFC-12 at Mace Head, Ireland**

*Notes*: The upper curve shows the model predicted concentrations, while the lower show the observations in year 2000. The observations are plotted as negative values, i.e. a concentrations of -5 pptv (parts per trillion) means a concentration of 5 pptv. The figure is prepared by Kjersti Ellingsen, University of Oslo.

For use in the Kyoto Protocol, inversion-modelling on a national scale seems most promising for $SF_6$, PFCs and HFCs, since there is no natural contribution to **F**, and since the loss (**Q**) takes place in the upper atmosphere (for $CH_4$ the loss is mainly at low altitude through reaction with hydroxyl (OH) radicals) and hence does not impose large fluctuations in the surface concentrations. Thus for the most important GHGs ($CO_2$, $CH_4$ and $N_2O$), the verification process has to be based on reported emission inventories with independent review.

## Discussion

When discussing methods to verify whether parties to the Kyoto Protocol are in compliance, it is important to distinguish between legal compliance and scientific compliance, as defined in the introduction to this chapter.

Due to this distinction, a party which has a commitment to reduce its emissions can be in compliance with the Protocol in legal terms, even if its real emissions exceed its Protocol commitment period target. Even if a party applies a set of measures, reports according to an emission inventory prepared according to the rules of the Kyoto Protocol (i.e., following the good practice guidelines of the IPCC (2000)), and has that report reviewed by an independent ERT, there remains the very real possibility that, due to incomplete scientific knowledge behind the relevant guidelines, that party's real emissions may not be in compliance with the party's legal commitments. For parties who choose to implement abatement measures on source categories where the emission factors are uncertain (e.g. agriculture or managed ecosystems), this could be a problem.

During the first commitment period of the Kyoto Protocol, the main task of verification will be to verify trends between 1990 and 2008–12. Generally, estimates of trends are less uncertain than estimates of absolute emissions, since some of the biases will be equal for both periods. For both methods discussed above, the base year emissions are the most uncertain. For reported national inventories, the input data needed to make a detailed inventory might be missing (see Figure 4.2), or the emission factors used for the 1990 estimate may be uncertain and difficult to improve as practices/technologies have changed. For the current commitment period, it has been estimated that the $2\sigma$ uncertainty in the trend estimates is about 4 per cent, which is of equal magnitude to common commitments taken on by the parties. Figure 4.7 shows an example of a party that is required to reduce emissions by 5.2 per cent under the Kyoto Protocol, with associated uncertainties in the trend estimates. Formally, uncertainties should not play a role (only if the guidelines are followed properly) when the national emission inventory is finally approved. However, a party has the right to defend itself before the Compliance Committee, and a potentially important question is what level of confidence is acceptable in applying sanctions to a party that appears to be in non-compliance with the Protocol. This question is directly linked to the standard legal principle that the defendant should be proved guilty beyond any reasonable doubt. If a confidence level of 5 per cent ($2\sigma$) is adopted, a look at Figure 4.7 from the perspective of the Enforcement Branch means that if the party reports an emissions reduction of only 1 per cent, it would be off the hook, since there is a 2.5 per cent probability that the reductions are 5.2 per cent or larger. A confidence level of 0.1 per cent ($4\sigma$) would let that same party off the hook even with a reported *increase* of 3.3 per cent (upper dashed line in Figure 4.7). To what extent the treatment of uncertainties will become an issue in the compliance regime of the Kyoto Protocol will be seen at the end of the first commitment period. However, in terms of assessing the effectiveness of the Kyoto Protocol (i.e., whether it has really achieved the reduction in the emissions agreed upon), it is nevertheless necessary to keep in mind the uncertainties depicted in Figure 4.7.

For IM, it is problematic that a proper network is lacking for 1990 atmospheric observations. For the next commitment period (2012–17) the situation will hopefully be different, since we would expect that at that time we will have better emissions data for the start year and that proper networks for observations of GHG concentrations can be put in place.

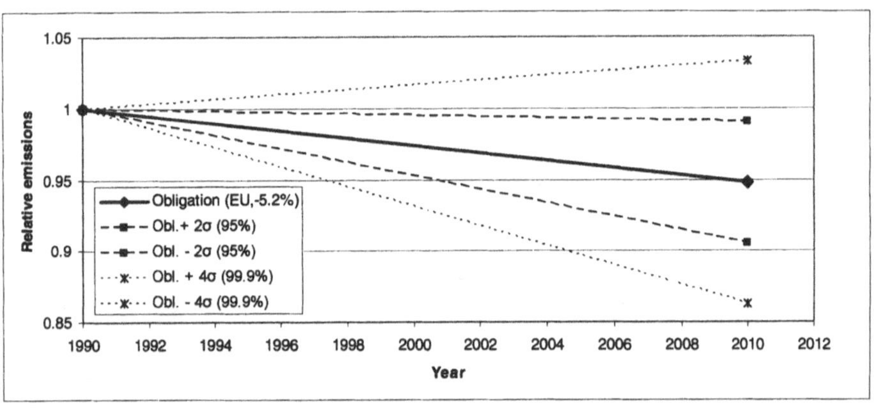


**Figure 4.7 Schematic illustration of targets and 2σ (dashed lines) and 4σ (dotted lines) confidence intervals under the Kyoto Protocol**

*Note*: σ = standard diviation

For the current commitment period under the Kyoto Protocol, we believe that it is highly unlikely that emission estimates based solely on IM can be used to impose penalties. However, we propose that a combination of the reporting and IM methods can be applied. We suggest that that if IM indicates that a party is not in compliance with the Protocol, a further international review of that party's reported national inventories is required. The additional review should aim at revealing whether the party has undertaken measures that, according to the IPCC guidelines, should have resulted in the appropriate emission reductions; or whether the measures undertaken are not sufficient according to the guidelines. In the first case, the discrepancy must be attributed to shortcomings in the scientific basis for the IPCC guidelines (errors in emission factors, unknown sources, etc.) or to errors in the IM, and the party should be deemed to be in compliance with the Protocol. In the latter case, the party must be deemed to be in non-compliance. As a side benefit, verification that includes the use of IM methods can help to improve the guidelines for future commitment periods.

Whether or not IM methods will be used more extensively for verification during future commitment periods (beyond 2008–12) will depend on several factors. We have mainly focused on the limitations imposed by uncertainties, but also of significant importance is the experience that will be gained during the first commitment period. If there is a widespread suspicion that some parties are cheating by underreporting emissions, there might be a stronger political demand for more independent verification. Without a political demand, the application of IM methods is unlikely, independent of potential reductions in the uncertainties of inverse methods through improved measurement network techniques and improved transport models. However, it must be kept in mind that a renegotiation of the Kyoto Protocol would be needed in order to change the verification system. This could be a significant obstacle even with a political demand from some parties, since the renegotiation of one part of the Protocol could open up other parts for renegotiation, potentially jeopardizing the whole agreement.

## Notes

The work undertaken at CICERO and NILU has been supported by a grant from the Norwegian Research Council (SAMSTEMT). We would like to thank Kristin Rypdal and Lynn Nygaard for their valuable comments and suggestions.

1 Kyoto Protocol, Art 3.1 and Annex A.
2 Kyoto Protocol, Art 8.
3 Troposphere: the part of the atmosphere extending from the surface to about 10km in altitude in mid-latitudes (ranging from 9km at high latitudes to 16km in the tropics).
4 The Intergovernmental Panel on Climate Change (IPCC) (see, e.g., IPCC, 1996a) gives this definition of radiative forcing (RF): the radiative forcing of the surface-troposphere system (due to a change, for example, in greenhouse gas concentration) is the change in net irradiance (in $Wm^{-2}$) at the tropopause after allowing stratospheric temperatures to re-adjust to radiative equilibrium, but with surface and tropospheric temperatures held fixed. RF is a metric for the strength of a climate change mechanism, and the main rationale for the use of RF can be illustrated by the simple equation that relates the global-mean RF to global-mean $\Delta T_s$ at equilibrium: $\Delta T_s = \lambda \cdot RF$, where $\lambda$ is a climate sensitivity parameter. Calculations with different climate models lead to a range of 0.4 – 1.2 $K(Wm^{-2})^{-1}$ for $\lambda$, reflecting the uncertainties in the response of the climate system.
5 Stratosphere: the region of the atmosphere above the troposphere up to about 50km.
6 Global warming potentials (GWPs) are a relative measure and are defined as the time integrated radiative forcing from the instantaneous release of 1kg of a trace gas i expressed relative to that of 1kg of the reference gas $CO_2$.
7 www.ipcc-nggip.iges.or.jp/public/gp/gpgaum.htm
8 Kyoto Protocol, Art 8.
9 The uncertainties are estimated by each country, and do not necessarily reflect real differences in uncertainties.
10 Kyoto Protocol, Art 8.3.
11 cf the RECAB project, www.biotheon.com/recab
12 www.nilu.no/niluweb/services/soge

## References

Bousquet, P., Ciaia, P., Peylin, P., Ramonet, M. and Monfrey, P. (1999) 'Inverse Modelling of Annual Atmospheric $CO_2$ Sources and Sinks. Part 1: Method and Control Inversion', *Journal of Geophysical Research*, vol 104, 26161–93

Geller, L. S. et al. (1997) 'Tropospheric $SF_6$: Observed latitudinal distribution and trends, derived emissions and interhemispheric exchange time', *Geophysical Research Letters*, vol 24, no 6, 675–8

Gurney, K. R., Law, R. M., Denning, A. S., Rayner, P. J., Baker, D., Bousquet, P., Bruhwiler, L., Chen, Y. H., Ciais, P., Fan, S., Fung, I. Y., Gloor, M.,

Heimann, M., Higuchi, K., John, J., Maki, T., Maksyutov, S., Masarie, K., Peylin, P., Prather, M., Pak, B. C., Randerson, J., Sarmiento, J., Taguchi, S., Takahashi, T. and Yuen, C. W. (2002) 'Towards Robust Regional Estimates of $CO_2$ Sources and Sinks Using Atmospheric Transport Models', *Nature*, vol 415, 626–9

Houweling, S., Kaminski, T., Dentener, F., Lelieveld, J. and Heimann, M. (1999) 'Inverse Modeling of Methane Sources and Sinks Using the Adjoint of a Global Transport Model', *Journal of Geophysical Research*, vol 104, 26137–60

Höhne, N. and Harnisch, J. (2002) 'Comparison of Emission Estimates Derived from Atmospheric Measurements with National Estimates of HFCs, PFCs and $SF_6$', paper presented to the Third International Symposium on Non-$CO_2$ Greenhouse Gases (NCGG-3), Maastricht, the Netherlands, 21–23 January

IPCC (1996a) *Climate Change 1995: The Science of Climate Change*, Cambridge University Press, Cambridge

IPCC (1996b) *IPCC Guidelines for National Greenhouse Gas Inventories*, vols 1–3, Intergovernmental Panel on Climate Change, London

IPCC (2000) *Emissions Scenarios, 2000: Special Report of the Intergovernmental Panel on Climate Change*, Cambridge University Press, Cambridge

IPCC (2001) *Climate Change 2001*: *The Scientific Basis*, Cambridge University Press, Cambridge

Kaminski, T., Heimann, M. and Giering, R. (1999) 'A Coarse Grid Three-Dimensional Global Inverse Model of the Atmospheric Transport, 2: Inversion of the Transport of $CO_2$ in the 1980s', *Journal of Geophysical Research*, vol 104, 18555–81

Maiss, M. and Brenninkmeijer, C. A. M. (1998) 'Atmospheric $SF_6$, trends, sources and prospects', *Environmental Science and Technology*, vol 32, no 20, 3077–86

Maiss, M. and Brenninkmeijer, C. A. M. (2000) 'A reversed trend in emissions of $SF_6$ into the atmosphere?' in Ham, J. van et al (eds) *Non-CO2 Greenhouse gases: Scientific Understanding, Control and Implementation*, Kluwer Publishers, Dordrecht, 199–204

Oram, D. E. et al. (1996) 'Recent tropospheric growth rate and distribution of HFC-134a ($CF_3CH_2F$)', *Geophysical Research Letters*, vol 23, no 15, 1949–52

Penman, J., Kruger, D., Galbally, I., Hiraishi, T., Nyenzi, B., Emmanuel, S., Buendia, L., Hoppaus, R., Martinsen, T., Meijer, J., Miwa, K. and Tanabe, K. (eds 2000) *Good Practice Guidance and Uncertainty Management in National Greenhouse Gas Inventories*, IPCC/OECD/IEA/IGES, Hayama, Japan

Rypdal, K. and Winiwarter, W. (2001) 'Uncertainties in Greenhouse Gas Emission Inventories: Evaluation, Comparability and Implications', *Environment Science and Policy*, vol 4, 107–16

Simmonds, P. G. et al. (1998) 'Calculated trends and the atmospheric abundance of 1,1,1,2-tetrafluoroethane, 1,1-dichloro-1-fluoroethane, and 1-chloro-1,1-difluoroethane using automated in-situ gas chromatography-mass spectrometry measurements recorded at Mace Head, Ireland from October 1994 to March 1997', *Journal of Geophysical Research*, vol 103, 16029–38

Winiwarter, W. and Rypdal, K. (2002) 'Assessing the Uncertainty Associated with a National Greenhouse Gas Emission Inventory: A Case Study for Austria', *Atmospheric Evironment*, vol 35, 5425–40

Zander, R. et al. (2000) 'Long-term evolution of the loading of $CH_4$, $N_2O$, CO, $CClF_2$, $CHClF_2$ and $SF_6$ above Central Europe during the last 15 years in Ham, J. van et al. (eds) *Non-CO2 Greenhouse gases: Scientific Understanding, Control and Implementation*, Kluwer Publishers, Dordrecht, 211–16

Chapter 5

# Effective Enforcement and Double-edged Deterrents: How the Impacts of Sanctions also Affect Complying Parties

Cathrine Hagem and Hege Westskog

## Introduction

At COP-7, in Marrakesh, the parties to the Kyoto Protocol agreed to establish a Compliance Committee with Facilitative and Enforcement Branches (UNFCCC, 2001). The Enforcement Branch will serve as a forum for determining whether Annex B parties have met their emissions targets and complied with their monitoring and reporting requirements (see Chapter 2 by Ulfstein and Werksman). It consists of ten members, drawn from countries that are parties to the Protocol, who will serve for no more than two consecutive four-year terms. If the Enforcement Branch determines that a party has exceeded its emission target under Article 3.1, it is required to apply the following sanctions:[1]

1 Deduct from the party's assigned amount for the second commitment period a number of tonnes equal to 1.3 times the number of tonnes of excess emissions in the first commitment period.
2 Require the development of a Compliance Action Plan by the non-compliant party.
3 Suspend the non-compliant party's eligibility to make transfers[2] under Article 17.[3]

The purpose of this chapter is twofold: first, to illustrate how the sanctioning of non-complying parties itself imposes costs and creates benefits for complying parties; and second, to consider how the composition of the Enforcement Branch may affect decisions to impose sanctions in light of these costs and benefits, if the members of the Enforcement Branch act in the interests of their own countries rather than in their individual capacities. We only intend to show the weakness of the sanction mechanism with regard to the possibility of the members of the Enforcement Branch to take their own countries' interests into account when making decisions, and how this may influence the decisions of the Branch. This implies that our argument will be of interest as long as the sanctions mechanisms are carried out at least once, and as long as some members of the Branch would take own interests into account when making decisions. We do not discuss the rationale for parties' non-compliance, neither do we argue that all the members of the Enforcement Branch will in fact use their own countries' interests as a basis for making decisions.

The degree to which members of the Enforcement Branch have an opportunity to take their own countries' economic interests into account in carrying out their responsibilities within the Compliance Committee is subject to debate. At first glance, there seems to be no room for dispute over the issue of non-compliance, and hence no room for the exercise of discretion. Each Annex B Party has agreed to an assigned amount before the start of the first commitment period. Emissions and transactions are recorded and registered by the party and assessed by an Expert Review Team (ERT).[4] At the end of the commitment period, each Annex B party's total emissions will be compared to its assigned amount and its use of the flexibility mechanisms. However, as Rypdal et al (2003) argue, this procedure masks an exercise that necessarily requires judgement and discretion, particularly with regard to uncertainties related to the calculation, assessment and adjustment of national inventories of emissions (see chapter 4 by Berntsen and associates, and Chapter 5 by Mitchell). Hence, it is not always obvious whether a country is in compliance or not, and the Enforcement Branch may have to use a considerable amount of discretion in reaching its determinations. However, once the issue of compliance is settled, the Marrakesh Accords leave no room for discretion about the form the sanctions will take (see Chapter 2 by Werksman and Ulfstein). In this chapter we take it as a given that there is some room for discretion by the Enforcement Branch members, and hence this discretion affords members an opportunity to take their own country's economic interests into account. We use the expression 'impose sanctions' to refer to the consequences of an Enforcement Branch decision that a country is in non-compliance with its Article 3.1 commitment.

Further, even when it is possible for Enforcement Branch members to take their own countries' interests into account, it might be questioned whether they would actually do so. First, the Marrakesh Accords expressly provide that members of the Branch shall serve 'in their individual capacity'. This implies that the members should not take instructions from their country of nationality or from any other country. On the other hand, the composition of the Branch is based on a specific distribution of the members across geographical regions and between Annex I and non-Annex I parties. This suggests that the Conference of the Parties (COP) acknowledges the usefulness of involving representatives from various countries with potentially differing economic interests in both the determination of compliance, and the implementation of sanctions for non-compliance. If it could be guaranteed that the members of the Enforcement Branch served only in their individual capacity, this complex composition of the Enforcement Branch with respect to the members' nationality would seem unnecessary (and even counterproductive, since the best qualified candidates are not necessarily chosen; see Chapter 2 by Ulfstein and Werksman). The discussion in this chapter proceeds under the assumption that some Enforcement Branch members may take their own country's interests into account when taking decisions. When this occurs, the composition of the Enforcement

Branch could influence whether sanctions are imposed on a country considered for non-compliance.

## Impacts of sanctions on complying and non-complying states

The features of the three different sanction mechanisms to be applied for failure to meet Article 3.1 commitments differ with respect to their impacts on non-complying and complying countries.[5] Implementation of the first sanction, the deduction of tonnes from the second commitment period assigned amount, can be viewed as a penalty directly related to the degree of non-compliance. Assuming the non-complying party would otherwise be eligible to participate in emissions trading in the second commitment period, the penalty per unit of excess emission in the first commitment period will correspond to the international quota[6] price in the second period multiplied by 1.3, and this penalty per unit of excess emission would be identical for all non-complying parties. If the non-complying parties' aggregate over-all excess emissions in the first commitment period are small, this sanction, imposed on parties found to be in non-compliance, will have little effect on complying parties' own costs of meeting their commitments.[7] However, if non-complying parties' excess emissions are large, the cost of compliance for other parties could be affected through impacts on the market price of quotas in the second commitment period.

The second sanction mechanism, development of a Compliance Action Plan, can be more costly for some non-complying countries than others, depending on the structure of their economies. However, the cost of this sanction mechanism will probably be of minor importance for non-complying parties. This sanction will surely have no impact on complying countries' costs of fulfilling their own obligations.

With respect to the third sanction mechanism, the economic cost of suspension from eligibility to sell quotas under Article 17 may differ considerably between countries, depending on their domestic abatement costs. Non-compliant countries that expected to benefit significantly from quota sales in the second commitment period would lose more than non-compliant countries that expected to sell small or no amounts of quotas. Non-complying countries that expect to be buyers of quotas will not be affected by this sanction mechanism, since it only affects a country's eligibility to sell. The suspension of a country from eligibility to sell quotas may also have an impact on the cost of compliance for other Kyoto parties, through alteration of the international price of quotas and through changes in the international prices of emission-intensive goods and fossil fuels. A suspension of non-compliant parties' eligibility to sell may drive up the international quota price. Complying countries that are net sellers will benefit from a higher international quota price, while countries that are net buyers of quotas will lose. Hence, both the first and the third sanction mechanisms may affect the cost of compliance for complying parties. This may give incentives for strategic considerations by the enforcing countries.

Axelrod and Keohane (1985) identify three sanctioning problems, each of which is relevant here:

1 The inability to identify non-compliers, because determining whether or not a country is in non-compliance requires judgement and discretion.

2 The inability to focus retaliation on non-compliers, as the suspension of eligibility to sell quotas has implications not just for the non-compliant party, but also for other parties.
3 The lack of incentive to sanction non-compliers, due to the potentially double-edged nature of the sanctions imposed.

At any point in time, ten countries will have one member each in the Enforcement Branch, with responsibility for determining whether parties are in compliance with their commitments and applying sanctions if they are not.[8] After a suspension of eligibility has been applied to a party, the Enforcement Branch can reinstate eligibility on the request of the suspended party, based on its submission of a Compliance Action Plan, unless the Branch determines that the party has not demonstrated that it will meet its target in the subsequent commitment period.[9] Hence, the representatives of the Enforcement Branch are in a position to make decisions with respect to sanctions on parties considered for non-compliance that may have important economic consequences for their own countries, and this may undermine the deterrent effect of the compliance system as a whole. At the same time, if a country heading for non-compliance perceives that it is in the economic interest of the members of the Enforcement Branch to allow it to sell quotas in the second commitment period, the deterrent effect of the suspension is weakened. That party may question whether sanctions will be implemented at all, and assume that even if implemented, suspension from the sale of quotas may be only temporary.

## The effect of sanctions on prices of quotas, fossil fuels and emission-intensive goods

A country that is sanctioned for non-compliance with Article 3.1 commitments under the Kyoto Protocol could respond in different ways. It could opt to withdraw from the agreement (avoiding sanctions and its commitments under the Protocol), partly comply with the sanctions imposed and the Kyoto commitment in the second commitment period,[10] or bring itself back into full compliance with its commitments and imposed sanctions in the second commitment period. Which strategy a state will adopt will be influenced by the expected costs and benefits of compliance. A party would presumably consider withdrawal from the agreement if the expected costs of compliance are higher than (or at most equal to) the benefits it expects to receive from fulfilling its obligations.[11] The expected costs of continuing to participate in the agreement will be larger when the party is sanctioned for non-compliance. Hence, a party found to be in non-compliance could find it more advantageous than before to withdraw from the agreement because of the increased costs of participating that result from the sanctions.

The imposition of sanctions on a country found to be in non-compliance could affect the prices of quotas, fossil fuels and emission-intensive goods[12] regardless of which strategy the non-compliant party adopts – that is, whether it chooses to withdraw or comply – but the price effects would differ. We would expect that the degree to which sanctioning affects prices would depend on the non-compliant country's expected volume of quota sales and its balance of export and import of fossil fuel products and emission-intensive goods.

First, with respect to volume of quota sales, if the non-compliant country expects to be a large seller of quotas and its eligibility to trade is suspended, the international quota price will increase with either the country's withdrawal from the agreement or the party's cooperation with sanctions. Upon withdrawal from the Protocol, the non-compliant country loses eligibility to participate in emissions trading. This will result in an increased quota price as a result of the reduced supply of quotas. On the other hand, if the same country chooses to comply with its commitments and sanctions, the quota price will increase due to the suspension of its eligibility to sell quotas.

If the non-complying country is a large buyer of quotas and withdraws, the quota price will decrease as a result of lower demand for quotas. If the non-complying net buyer instead chooses to comply with sanctions in the second commitment period, it will face a lower assigned amount as the result of a deduction of tonnes, which will require a greater emission-reduction commitment and hence lead to an increased quota demand. This, in turn, will increase the international quota price.

The volume of quota trade will be determined by domestic abatement costs and parties' negotiated emission-reduction commitments. An Annex B party's negotiated emission reduction commitments for the first commitment period will be the difference between its business-as-usual emissions (BAU emissions)[13] and the initial number of quotas allocated to it through assigned amounts under the Kyoto Protocol. The lower the BAU emissions of a party, the greater the initial amount of quotas allocated to it, and the lower abatement costs, the higher the amount of quotas that country offers on the international quota trading market. Hence, the greater the impact on quota prices should that country be found to be in non-compliance and choose to either withdraw or comply through the acceptance of sanctions.

Several studies have analysed Annex B trading under the Kyoto Protocol. The general picture is as we would expect: under the Kyoto Protocol's trading regime, Russia and other former Soviet Union and Eastern European countries would become major sellers of quotas, and the Organization for Economic Cooperation and Development (OECD) countries would be net purchasers of quotas. See, for instance, Nordhaus and Boyer, 1999; Bollen, Gielen and Timmer, 1999 and Cooper et al, 1999.[14] However, some OECD countries would become sellers of quotas. A study by Holtsmark and Mæstad (2002) shows that Greece, Ireland, Portugal, Spain and Australia would be quota sellers under the Kyoto Protocol, whereas Italy, the UK, Sweden, the Netherlands and Germany will be buyers.

Second, the pattern of export or import of various fossil fuel products could be of importance for the price effects of the sanctions on these products. If the non-complying country chooses to withdraw from the agreement, its consumption of fossil fuels may increase, or decrease more slowly, if it considers itself no longer bound by its Kyoto emission-reduction commitment. If the country has a more than an insignificant share of total world consumption of specific fossil fuels, increasing consumption may increase the price of these fossil fuels. If, instead, the non-complying country chooses to remain within the Kyoto framework, this will have three effects:

1 If the country had expected to be a large seller of quotas due to relatively low abatement costs, the suspension of its eligibility to sell quotas would remove this incentive for domestic abatement efforts. This might imply fewer emission-reduction measures than would otherwise have occurred. On the other hand, this would imply that other complying countries will have to do more at home.[15] The price pattern between the different fossil fuel products could be changed as a result of the non-complying party's suspension from eligibility to sell quotas. The non-compliant country would be likely to increase its consumption of fossil fuels, whereas the compliant countries would decrease theirs. If countries' consumption patterns of the various fossil fuels, for instance their relative reliance on natural gas, differ between the non-compliant country and the compliant countries, this would change the price differential between the various fossil fuels.
2 The first sanction mechanism (where the assigned amount for the second commitment period is reduced by an amount equal to the excess emissions in the first period multiplied by 1.3) could also affect the price of fossil fuels if a country with significant excess emissions is sanctioned, as the total agreed emission reductions for the first commitment period are effectively increased due to the use of a multiplier.
3 Third, the price of various emission-intensive goods could also be influenced by sanctioning a non-compliant country. If the non-compliant country chooses to withdraw from the Protocol, its production of emission-intensive goods could increase. If the non-compliant country is responsible for a substantial share of the total world production of these goods, prices for these goods may decrease. If, instead, the non-complying country chooses to comply with the sanctions, the suspension of its eligibility to sell quotas will not significantly change the total amount of carbon emitted under the Protocol. However, the export and import pattern of different emission-intensive goods might be altered, with consequent impacts on product pricing, if a large quota seller is suspended from quota trade. The resulting change in price differentials between various emission-intensive products will depend on the export and import product mixes of the non-complying country and the other parties. Also, as for fossil fuels, the price of emission-intensive goods could be altered as a consequence of imposing the 1.3 penalty on each unit of excess emissions.

## How sanctioning a non-complying party can have economic consequences for enforcing countries

The previous section showed how sanctions could affect prices of quotas, fossil fuels and emission-intensive goods. Because countries in the Enforcement Branch (referred to here as 'enforcing countries') also participate in the global economy, these impacts could affect their costs and benefits of continued participation in the agreement. More importantly, these impacts could influence their decisions. The extent to which suspending a non-complying country's eligibility to sell quotas and imposing a penalty on its excess emissions will have consequences for an enforcing country depends on whether the enforcing country:

1 is a seller of quotas/Clean Development Mechanism (CDM) credits or a buyer of quotas;
2 has a large emission-intensive industry/is a significant importer of emission-intensive goods; and
3 is a fossil-fuel importer or exporter.

First, changes in the international price of tradeable quotas would influence the expected costs and benefits of an enforcing country differently depending on whether that country is a seller or a buyer of tradeable quotas. If the international quota price increases as a result of imposing sanctions on a non-complying country, this would be an advantage for a selling country but a disadvantage for a buying country, and vice versa if the international quota price decreases. Whether a country ends up as a buyer or seller of quotas in the international market depends, as discussed above, on the abatement costs of that country and its emission-reduction commitments under the Kyoto Protocol.

Developing countries will be affected by changes in the quota prices because they may participate indirectly in the tradeable quota markets by selling CDM credits. The tradeable quota market and the market for CDM credits will be linked by the international quota price. The price of a tradeable quota and a CDM credit will probably be nearly equal (including transaction costs).[16] Large sellers of CDM credits will benefit from an increase in the prices of quotas. China and India emerge as major CDM suppliers according to both the Zhang (2000) study and a study by Austin and Faeth (2000). According to Zhang, China and India will account for, respectively, 60 per cent and 16 per cent of the total CDM flows. In the Austin and Faeth study, China's share of the CDM flow ranges from 57 to 70 per cent, while India accounts for 7 to 11 per cent.

Second, if the prices of different emission-intensive goods are changed, this will affect the costs and benefits of enforcing countries depending on their magnitude of import and export of these goods. A large exporter of specific emission-intensive goods would benefit from a price increase, whereas the opposite would be the case for a large importer of the same type of goods. An enforcing country that imports and exports a small amount of those goods would not be significantly affected by the price changes.

Finally, the same would be the case for a large exporter or importer of different fossil fuels and for a large producer or importer of different emission-intensive goods. Consider Tables 5.2 and 5.3 in Appendix A: such large oil exporters (relative to GDP) as the Middle East countries, Venezuela, Norway (in the category 'rest of EFTA') and the whole category 'rest of the sub-Saharan African countries' would benefit from an increase in the price of oil. Large oil importers (relative to GDP), such as Korea, the Philippines, Thailand, and Central America and the Caribbean would experience an increase in costs from an increase in the price of oil.

## The composition of the Enforcement Branch: An example

Whether or not the factors elaborated above will influence the decision making of the Enforcement Branch as a whole depends largely on the composition of the Enforcement Branch. This section presents examples of different compositions of the Enforcement Branch that may occur under the specific rules for the regional distributions of the members. We show that under the voting regulations for the Enforcement Branch, one specific composition of the Enforcement Branch may lead to a decision to impose sanctions for non-compliance, while another composition of the branch may not.

According to the Marrakesh Accords, the Enforcement Branch consists of ten members. as follows:[17]

1 one member from each of the five regional groups of the United Nations,[18] and one member from the small island developing states (SIDS), taking into account the interest groups as reflected by the current practice in the Bureau of the Conference of the Parties;
2 two members from parties included in Annex I; and
3 two members from parties not included in Annex I.

Voting rules for the Enforcement Branch and the Compliance Committee are also stipulated.[19] The decisions of the Compliance Committee may be made by a majority of at least three-quarters of the members present and voting. In addition, the adoption of decisions by the Enforcement Branch requires a majority of members from parties included in Annex I present and voting, as well as a majority of members from parties not included in Annex I present and voting.

Lets us consider two different compositions of the Enforcement Branch, both satisfying the rules referred to above.

**Table 5.1 Alternative compositions of the Enforcement Branch**

| | |
|---|---|
| A | 1) Tanzania, China, Poland, Bolivia, Australia and the Maldives |
| | 2) Japan and Italy |
| | 3) India and Egypt |
| B | 1) Tanzania, China, Poland, Bolivia, Australia and the Maldives |
| | 2) Japan and Spain |
| | 3) India and Egypt |

In both compositions there are four members from parties included in Annex I and six members from parties not included in Annex I. The only difference between the two compositions is that a member from Spain replaces a member from Italy.

As discussed in the previous section, whether a complying country will gain or lose when a non-complying country faces sanctions depends on a number of complex world market effects that pull in different directions. Hence, it is not possible for any enforcing country to have complete knowledge about the economic impact of voting in favour of an implementation of the sanctions. However, the extensive literature on emission trading (see this chapter's third section) suggests that it is likely that an enforcing country will have a general idea about how sanctions may affect quota prices and whether a low or a high quota price is beneficial for its economy.

We assume in this section that members of the Enforcement Branch act in the interest of their country of nationality. Since the purpose of our example is to show that the composition of the Enforcement Branch may influence the decisions, we simplify the analysis by making certain assumptions regarding the enforcing countries' interests. First, we assume that the enforcing countries are concerned about the economic impacts of changes in the quota price resulting from imposing sanctions.[20] Second, we assume that enforcing countries prefer to impose sanctions if they do not have a negative economic impact. If the sanctions are likely to result in negative economic impacts for the enforcing country, we assume that they would prefer not to impose sanctions.

We consider a hypothetical situation where considerable evidence suggests that Russia is in non-compliance. As discussed above, several empirical models show that Russia, like other East European countries, is likely to be a large seller of quotas. If Russia loses its eligibility to sell quotas, the international quota price will rise significantly. The rise in permit prices will occur whether Russia responds to the sanctions by withdrawing from the agreement or not. An increase in the permit price will surely hurt buyers of quotas, and be beneficial for other sellers of quotas and non-Annex I countries with a large potential for hosting CDM projects.

In our example, at least two of the non-Annex I countries have a potential for hosting inexpensive CDM projects (China and India), while the other non-Annex I countries may be indifferent to changes in the international quota prices. Hence, all the non-Annex I countries in our example would prefer that sanctions were imposed.

According to a numerical model study by Holtsmark and Mæstad (2002), Italy and especially Japan will be large buyers of quotas under the Kyoto agreement (with US participation).[21] Australia, Spain and especially Poland will be sellers of quotas. A higher quota price will be beneficial for Australia, Spain and Poland, and costly for Japan and Italy. On the assumptions above, Enforcement Branch members from the non-Annex I countries and Australia, Spain and Poland will vote in favour of sanctioning Russia, while the members from Japan and Italy will vote against. We see from our example that composition *A* of the Enforcement Branch will not reach a decision to declare Russia in non-compliance and implement the sanctions, while a replacement of the representative from Italy with a representative from Spain (composition *B*) would lead to an implementation of the sanctions.

Under composition *A*, the requirement of a majority among the Annex I countries blocks the decision of non-compliance, although more than three-quarters of the members favour implementing the sanctions.

In our example, we assumed that enforcing countries are only concerned about the economic impact following from a possible implementation of the sanctions. Obviously, there are concerns other than the pure economic effects on the national economy that may also be taken into account when enforcing countries evaluate whether they prefer the sanctions to be implemented. Enforcing countries may have strong preferences for preventing a weakening of the agreement. To carry out the sanctions when non-compliance is observed will surely strengthen the deterrence effect. On the other hand, sanctions may lead to withdrawal from the agreement and thus weaken the environmental effect of the agreement. Enforcing countries' perceptions of the non-compliant countries' responses to sanctions are therefore relevant to their decisions of whether or not to impose sanctions.

## Concluding remarks

The main purpose of this chapter has been to elaborate on how differences in the economic structure among potential members of the Enforcement Branch lead to differences in the economic costs and benefits of sanctioning non-compliance. If a non-compliant country has a more than insignificant share of the total supply of quotas, suspending its eligibility to sell quotas will have impacts on the quota price and on the price differences between different fossil fuels and emission-intensive products. This will affect the economy of the enforcing countries. Imposing sanctions on a quota buyer may also affect the global economy if the country's excess emissions are large, because the sanctions will reduce the assigned amount for the second period. However, the imposition of sanctions on countries that deviate little from their emission commitments and did not expect to be large sellers of quotas would not affect the global economy, unless those countries withdraw from the agreement. If there is a possibility that a party found to be in non-compliance with its Article 3.1 commitments will withdraw from the agreement in response to the sanctions, the size of the sanctioned country's economy will influence the possible economic impact on the enforcing countries. A country that withdraws from the agreement no longer has to pay a price for the domestic emissions that result from the production of emission-intensive goods or the consumption of fossil fuels. The effect of a withdrawal on the international market for emission-intensive goods and energy grows with the market share of the sanctioned country.

All considered, it is clear that only when a small economy is considered for non-compliance can Enforcement Branch members be certain that they will not be economically affected. Since no enforcing country loses, the Enforcement Branch may easily reach consensus on imposing sanctions on small countries. When it comes to large non-complying countries, the members of the Enforcement Branch could have conflicting interests. As shown in the example in the preceding section, it is possible that sanctions will be imposed on a large non-complying country by one specific composition of the Enforcement Branch but not by another, if the members of the Enforcement Branch act in the economic interests of their own countries.

Another feature of the sanction mechanism that could be regarded as problematic is that sellers and buyers of quotas are treated unequally when considered for non-compliance. Since only sellers can be suspended from trading quotas, a non-compliance finding would have stronger economic consequences for sellers than for buyers. The latter could only be sanctioned in accordance with the first sanction mechanism.

Thus the deterrent effect of sanctions varies according to whether the non-complying country is a small or large quota seller, whether it is a seller or buyer of quotas, and the composition of the Enforcement Branch. The sanctions may have a moderate deterrent effect on both buyers of quotas in general and on large quota sellers, which see that imposing sanctions on them is not in the economic interests of the majority of the Enforcement Branch members.

When the members of the Enforcement Branch take national interests into account, it is worth exploring under which circumstances, and how exactly, they do so. Even if the interests of their own countries have an impact on the decisions of the members, these members might take into account more than just economic interests. A member of the Enforcement Branch could also consider the political interests of his or her country. For instance, if evidence of non-compliance is strong, an Enforcement Branch decision to the contrary could undermine the whole compliance system and the agreement itself. Also, a member could take into account the environmental damages associated with climate change. Suspending a non-complying country from selling quotas could lead the non-complying country to withdraw from the agreement and the damages from emissions of GHGs would increase, both because this specific country will do less, and because other countries might follow suit and withdraw as well.

The fundamental problem of the sanction mechanisms under the Kyoto Protocol is that what you can agree on behind the veil of ignorance could change when you get additional information on who is going to be sanctioned. Once the veil is lifted, members of the Enforcement Branch may behave strategically and impose sanctions not on the basis of the violation, but on the basis of how carrying out the sanctions may affect their own countries. This represents a significant potential weakness of the sanction mechanisms in the Kyoto Protocol.

Reducing the opportunity for strategic behaviour might strengthen the agreement. The current design of the sanction mechanisms is meant to serve two purposes: restitution and deterrence. By requiring that a non-complying country make up for its insufficient emission reductions in the next period, environmental restitution is made. Requiring that the country multiply the amount by which it was deficient by 1.3 serves as a deterrent. Deterrence is also ensured by giving an economic disincentive through the threat of revoking a country's eligibility to sell quotas on the international market. However, an equally effective economic deterrent – and one that would not encourage strategic behaviour on the part of the Enforcement Branch – could be to define a penalty per unit of exceeded emissions independent of quota trading. For quota sellers, this amount could be deducted from their income from quota sales by holding back a certain amount from quota sales until the end of the commitment period when possible non-compliance is revealed.

However, not even this would completely eliminate the possibilities for strategic voting among the members of the Enforcement Branch. The restitution element, that is, the reduction in the assigned amount corresponding to the excess emissions in the first period, increases global emissions reduction relative to a situation without sanctions. Increased global emissions reduction affects the prices of quotas, energy and emission-intensive goods, and would therefore affect countries differently. Hence, it is unavoidable that lifting the veil of ignorance might lead to strategic behaviour. Furthermore, since countries can withdraw from the agreement when they face sanctions, members of the Enforcement Branch can consider the costs and benefits of a possible withdrawal when they make their decisions.

Perhaps the most effective way to reduce strategic behaviour would be to take measures to ensure the independence of the members of the Enforcement Branch so that political and economic interests do not influence their decisions. For instance, when a seat of the Enforcement Branch is up for re-election, that member might take into account the interest of the electors when he or she decides whether to sanction non-compliance or not. Designing the selection process for the members of the

Enforcement Branch by relying more on legal qualifications and more permanent positions would reduce this strategic element for the members.[22]

## Notes

Comments from Jon Hovi and Lynn P. Nygaard are highly appreciated.

1 A Party facing non-compliance has the opportunity to continue to acquire emission reduction units (ERUs), certified emission reductions (CERs), assigned amount units (AAUs) and removal units (RMUs) until 100 days after the completion of the expert review process for the last year of the commitment period, in order to achieve compliance. See Marrakesh Accords, Decision 24/CP.7, Sec XIII. A Party that fails to cure its non-compliance during this 100-day 'true-up' period is subject to non-compliance consequences under Decision 24/CP.7. Also, the requirements of a commitment period reserve could reduce the risks of non-compliance, though not removing it completely. Parties must maintain 90 per cent of the assigned amount, or five times the most recently reviewed inventory as a commitment period reserve, whichever is the lower.

2 In this chapter, we use the term 'sale' to describe transfers of AAUs, ERUs, CERs and RMUs.

3 Art 17 of the Kyoto Protocol allows international emission trading between Annex B countries. Trade occurs through transfers and acquisitions of AAUs, ERUs, CERs and RMUs.

4 See Marrakesh Accords, Decision 23/CP.7

5 Note that all three sanction mechanisms will be applied to a party found to be in non-compliance. We discuss each sanction separately to better illustrate how sanctioning also affects complying parties, including the members of the Enforcement Branch.

6 We use the term 'quota' to refer to units that can be traded under Art 17 of the Kyoto Protocol, i.e. AAUs, ERUs, CERs and RMUs.

7 If the amount of excess emissions in the first period is small, the sanction is unlikely to influence the international quota price and hence unlikely to have implications for the cost of compliance for other parties.

8 The decision can be appealed to the COP, which may agree by a three-quarters majority vote of the parties present to override the decision and refer the matter of the appeal back to the Enforcement Branch (UNFCCC, 2001).

9 Marrakesh Accords, Decision 24/CP.7, Sec XV, para 5(c), Sec IX.

10 Partial compliance signifies here that a party to the agreement only partly fulfils its obligations to reduce emissions of GHGs.

11 Benefits are here defined widely. They include all kinds of benefits a party would receive from participating in the agreement, including benefits from reducing its own emissions and gaining political credibility. For most countries, the effects of reducing GHG emissions will mainly be connected to the co-benefits of reduced local air pollution (such as reduced damage to human health and environment from air pollutants: see, for instance, Aunan et al, 2004; Seip et al, 2001). However, for a country with a large share of the world's GHG emissions, reducing its own emissions by a significant amount may also have an impact on the rate of climate change.

12 The sanctions we are discussing here would also affect the prices of other goods through the market mechanisms. However, here we focus on those goods that will be most influenced by the sanctions.

13 BAU emissions (related to the Kyoto Protocol) are the emissions the parties would have in 2010 if emissions reductions were not undertaken

14 All these studies refer to a situation in which the US participates in the climate agreement. The picture prevails also without the US's participation when it comes to who buys and who sells quotas. See the study by Böhringer, 2002.

15 The emission-reduction requirements agreed under the Protocol should still be fulfilled by the complying countries, even though a reallocation of emission reductions could imply a different composition of the GHGs reduced.

16 An Annex B party will not be willing to buy CDM credits if those credits are more expensive than tradeable quotas on the international quota market. Hence, the difference in prices between those two mechanisms will probably approach zero.

17 Marrakesh Accords, Decision 24/CP.7, Sec V, para 1.

18 The five regional groups are African states, Asian states, East European states, Latin American and Caribbean states, and Western European and other states.

19 Marrakesh Accords, Decision 24/CP.7, Sec II, para 9.

20 We do not take into consideration the effects on prices of energy and emission-intensive goods in this example. In a forthcoming study (Hagem et al, 2004), a detailed empirical general equilibrium model is used to calculate the economic impact of sanctioning non-compliance. In this study, all the effects on equilibrium prices in the energy market and emission-intensive goods markets are included in the calculations of how countries are affected by sanctioning non-complying countries, both by suspending them from eligibility to sell quotas and imposing a penalty per unit of excess emissions.

21 In their study, the internal EU distribution of the Kyoto commitment is the starting point for the distribution of commitments across EU members.

22 See Chapter 2 by Ulfstein and Werksman for a further discussion of this issue.

## References

Aunan, K., Fang, J., Vennemo, H., Oye, K. A. and Seip, H. M. (2004) 'Co-benefits of Climate Policy: Lessons Learned From a Study in Shanxi', *Energy Policy*, vol 32, no 4, 567–81

Austin, D. and Faeth, P. (2000) *Financing Sustainable Development with the Clean Development Mechanism*, World Resources Institute, Washington, DC

Axelrod, R. and Keohane, R. O. (1985) 'Achieving Cooperation under Anarchy: Strategies and Institutions', *World Politics*, vol 38, 226–54

Böhringer, C. (2002) 'Climate Politics From Kyoto to Bonn: From Little to Nothing?', *The Energy Journal*, vol 23, 51–73

Bollen, J., Gielen, A. and Timmer, H. (1999) 'Clubs, Ceilings and CDM: Macroeconomics of Compliance with the Kytoto Protocol', *The Energy Journal, Special Issue*: *The Costs of the Kyoto Protocol*: *A Multi-Model Evaluation*, vol 20, 177–206

Center for Global Trade Analysis (2001) *The GTAP5 Data Package – Global Trade, Assistance and Production*, CD-ROM, Purdue University, Indiana

Cooper, A., Livermore, S., Rossi, V., Wilson, A. and Walker, J. (1999) 'The Economic Implications of Reducing Carbon Emissions: A Cross-Country Quantitative Investigation using the Oxford Global Macroeconomic and Energy Model', *The Energy Journal, Special Issue: The Costs of the Kyoto Protocol: A Multi-Model Evaluation*, vol 20, 335–66

Hagem, C., Kallbekken, S., Mæstad, O. and Westskog, H. (in press 2004) 'Enforcing the Kyoto Protocol: Sanctions and Strategic Behavior', *Energy Policy*

Holtsmark, B. and Mæstad, O. (2002) 'Emission Trading under the Kyoto Protocol: Effects on Fossil Fuel Markets under Alternative Regimes', *Energy Policy*, vol 30, no 3, 207–18

Nordhaus, W. and Boyer, J. (1999) 'Requiem for Kyoto: An Economic Analysis of the Kyoto Protocol', *Energy Journal, Special Issue: The Costs of the Kyoto Protocol: A Multi-Model Evaluation*, vol 20, 93–130

Rypdal, K., Stordal, F., Fuglestvedt, J. S. and Berntsen, T. (2003) 'Assessing Compliance with the Protocol: Expert Reviews, Inverse Modeling or Both?', working paper 2003:7, CICERO, Oslo

Seip, H. M., Aunan, K., Vennemo, H. and Fang, J. (2001) 'Mitigating GHGs in Developing Countries', *Science*, vol 293, 2391–2

UNFCCC (2001) 'Marrakesh Accords', http://unfccc.int/cop7/

Zhang, Z. X. (2000) 'Estimating the Size of the Potential Market for the Kyoto Flexibility Mechanisms', *Weltwirtschaftliches Archiv, Review of World Economics*, vol 136, no 3, 491–521

# Appendix A: Tables

**Table 5.2 Bilateral imports, 1997**

| Country | % of total import, world prices | | | | Imports relative to GDP | | | |
|---|---|---|---|---|---|---|---|---|
| | Energy-intensive goods | Coal | Oil | Gas | Energy-intensive goods | Coal | Oil | Gas |
| Australia | 1.17 | 0.00 | 1.10 | 0.00 | 3.98 | 0.00 | 0.69 | 0.00 |
| New Zealand | 0.28 | 0.00 | 0.23 | 0.00 | 5.76 | 0.00 | 0.87 | 0.00 |
| China | 4.64 | 0.35 | 1.86 | 0.00 | 7.28 | 0.01 | 0.53 | 0.00 |
| Hong Kong | 1.46 | 0.97 | 0.00 | 0.52 | 14.02 | 0.17 | 0.00 | 0.18 |
| Japan | 4.91 | 25.56 | 12.10 | 13.20 | 1.55 | 0.15 | 0.70 | 0.15 |
| Korea | 2.93 | 9.10 | 6.43 | 3.36 | 8.80 | 0.51 | 3.55 | 0.37 |
| Taiwan | 1.97 | 6.48 | 1.75 | 0.82 | | | | |
| Indonesia | 0.99 | 0.06 | 0.46 | 0.00 | 6.35 | 0.01 | 0.54 | 0.00 |
| Malaysia | 1.23 | 0.40 | 0.13 | 0.00 | 15.53 | 0.09 | 0.31 | 0.00 |
| Philippines | 0.63 | 0.62 | 0.95 | 0.00 | 10.80 | 0.20 | 2.98 | 0.00 |
| Singapore | 1.61 | 0.00 | 2.95 | 0.03 | 27.01 | 0.00 | 9.08 | 0.02 |
| Thailand | 1.17 | 0.56 | 1.86 | 0.00 | 9.94 | 0.09 | 2.90 | 0.00 |
| Vietnam | 0.27 | 0.00 | 0.00 | 0.00 | 16.84 | 0.00 | 0.00 | 0.00 |
| Bangladesh | 0.14 | 0.00 | 0.04 | 0.00 | 4.19 | 0.00 | 0.22 | 0.00 |
| India | 1.09 | 1.70 | 1.72 | 0.00 | 3.66 | 0.11 | 1.06 | 0.00 |
| Sri Lanka | 0.07 | 0.00 | 0.12 | 0.00 | 6.38 | 0.00 | 1.85 | 0.00 |
| *Rest of South Asia* | 0.26 | 0.21 | 0.24 | 0.00 | 4.93 | 0.07 | 0.83 | 0.00 |
| Canada | 3.45 | 3.24 | 2.01 | 0.14 | 7.32 | 0.13 | 0.78 | 0.01 |
| US | 11.96 | 1.43 | 24.06 | 15.24 | 2.02 | 0.00 | 0.75 | 0.10 |
| Mexico | 1.85 | 0.41 | 0.00 | 0.16 | 6.37 | 0.03 | 0.00 | 0.02 |
| Central America and the Caribbean | 0.80 | 0.06 | 1.26 | 0.00 | 11.43 | 0.01 | 3.29 | 0.00 |
| Colombia | 0.36 | 0.00 | 0.00 | 0.00 | 5.10 | 0.00 | 0.00 | 0.00 |
| Peru | 0.17 | 0.08 | 0.22 | 0.00 | 3.53 | 0.03 | 0.82 | 0.00 |
| Venezuela | 0.31 | 0.06 | 0.00 | 0.00 | 4.96 | 0.02 | 0.00 | 0.00 |
| *Rest of Andean Pact* | 0.22 | 0.00 | 0.00 | 0.00 | 10.60 | 0.00 | 0.00 | 0.00 |
| Argentina | 0.67 | 0.21 | 0.05 | 0.22 | 2.76 | 0.02 | 0.04 | 0.03 |
| Brazil | 1.21 | 3.19 | 1.59 | 0.00 | 2.06 | 0.10 | 0.50 | 0.00 |
| Chile | 0.35 | 1.00 | 0.45 | 0.10 | 6.16 | 0.33 | 1.46 | 0.07 |
| Uruguay | 0.09 | 0.00 | 0.07 | 0.00 | 6.47 | 0.00 | 0.89 | 0.00 |
| *Rest of South America* | 0.09 | 0.00 | 0.01 | 0.00 | | | | |
| Austria | 1.39 | 0.95 | 0.49 | 1.12 | 9.43 | 0.12 | 0.61 | 0.28 |
| Belgium | 3.55 | 2.58 | 1.72 | 2.73 | 20.57 | 0.28 | 1.83 | 0.59 |
| Denmark | 0.87 | 1.98 | 0.28 | 0.00 | 7.10 | 0.30 | 0.41 | 0.00 |
| Finland | 0.66 | 1.36 | 0.55 | 0.62 | 7.54 | 0.29 | 1.16 | 0.27 |

| Country | % of total import, world prices | | | | Imports relative to GDP | | | |
|---|---|---|---|---|---|---|---|---|
| | Energy-intensive goods | Coal | Oil | Gas | Energy-intensive goods | Coal | Oil | Gas |
| France | 5.61 | 3.11 | 4.64 | 7.15 | 5.47 | 0.06 | 0.83 | 0.26 |
| Germany | 8.45 | 4.50 | 5.41 | 14.08 | 5.49 | 0.05 | 0.65 | 0.34 |
| UK | 5.54 | 4.11 | 2.63 | 0.16 | 5.76 | 0.08 | 0.50 | 0.01 |
| Greece | 0.57 | 0.22 | 0.97 | 0.02 | | | | |
| Ireland | 0.72 | 0.67 | 0.15 | 0.14 | 14.50 | 0.25 | 0.56 | 0.11 |
| Italy | 4.61 | 3.38 | 4.63 | 7.56 | 5.48 | 0.07 | 1.01 | 0.33 |
| Luxembourg | 0.16 | 0.10 | 0.00 | 0.15 | | | | |
| Netherlands | 3.19 | 2.77 | 3.22 | 1.04 | 11.92 | 0.19 | 2.21 | 0.14 |
| Portugal | 0.66 | 1.03 | 0.69 | 0.00 | 9.04 | 0.26 | 1.72 | 0.00 |
| Spain | 2.56 | 2.15 | 2.94 | 2.67 | 6.58 | 0.10 | 1.39 | 0.25 |
| Sweden | 1.30 | 0.72 | 1.03 | 0.01 | 7.63 | 0.08 | 1.11 | 0.00 |
| Switzerland | 1.85 | 0.03 | 0.27 | 0.49 | 9.96 | 0.00 | 0.26 | 0.10 |
| *Rest of EFTA* | 0.81 | 0.30 | 0.08 | 0.00 | 6.60 | 0.05 | 0.12 | 0.00 |
| Hungary | 0.47 | 0.45 | 0.29 | 1.41 | 15.34 | 0.27 | 1.72 | 1.71 |
| Poland | 0.97 | 0.62 | 0.80 | 1.39 | 10.41 | 0.12 | 1.59 | 0.56 |
| *Rest of Central and Eastern Europe* | 1.41 | 3.34 | 1.28 | 4.81 | 14.50 | 0.63 | 2.40 | 1.82 |
| Former Soviet Union | 1.92 | 4.51 | 1.77 | 18.13 | 4.38 | 0.19 | 0.74 | 1.53 |
| Turkey | 1.00 | 2.26 | 1.22 | 1.90 | 6.97 | 0.29 | 1.56 | 0.49 |
| *Rest of Middle East* | 2.24 | 1.59 | 1.38 | 0.00 | 5.76 | 0.08 | 0.65 | 0.00 |
| Morocco, Western Sahara | 0.17 | 0.50 | 0.28 | 0.00 | 6.39 | 0.35 | 2.00 | 0.00 |
| *Rest of North Africa* | 0.74 | 0.32 | 0.06 | 0.02 | 5.78 | 0.05 | 0.09 | 0.01 |
| Botswana | 0.04 | 0.00 | 0.01 | 0.00 | 12.48 | 0.00 | 0.36 | 0.00 |
| *Rest of South African Customs Union* | 0.54 | 0.10 | 0.62 | 0.00 | 5.20 | 0.02 | 1.09 | 0.00 |
| Malawi | 0.01 | 0.00 | 0.00 | 0.00 | 5.29 | 0.00 | 0.28 | 0.00 |
| Mozambique | 0.02 | 0.00 | 0.00 | 0.00 | 5.71 | 0.03 | 0.00 | 0.00 |
| Tanzania | 0.03 | 0.00 | 0.03 | 0.00 | 5.58 | 0.00 | 1.07 | 0.00 |
| Zambia | 0.02 | 0.00 | 0.03 | 0.00 | 6.62 | 0.00 | 1.66 | 0.00 |
| Zimbawe | 0.07 | 0.00 | 0.00 | 0.00 | 11.06 | 0.01 | 0.00 | 0.00 |
| *Rest of Southern Africa* | 0.07 | 0.00 | 0.01 | 0.00 | 7.15 | 0.00 | 0.10 | 0.00 |
| Uganda | 0.01 | 0.00 | 0.01 | 0.00 | 2.78 | 0.01 | 0.29 | 0.00 |
| *Rest of sub-Saharan Africa* | 0.59 | 0.50 | 0.46 | 0.00 | 5.04 | 0.08 | 0.72 | 0.00 |
| *Rest of World* | 0.82 | 0.17 | 0.44 | 0.59 | 4.02 | 0.02 | 0.39 | 0.11 |
| **Total** | **100.00** | **100.00** | **100.00** | **100.00** | | | | |

*Source*: Center for Global Trade Analysis (2001) (GTAP5)

*Note*: Blank cells indicate that no records are available.

**Table 5.3 Bilateral exports, 1997**

| Country | % of total export, world prices | | | | Export relative to GDP | | | |
|---|---|---|---|---|---|---|---|---|
| | Energy-intensive goods | Coal | Oil | Gas | Energy-intensive goods | Coal | Oil | Gas |
| Australia | 1.49 | 27.92 | 0.63 | 1.97 | 4.84 | 1.42 | 0.37 | 0.24 |
| New Zealand | 0.22 | 0.27 | 0.10 | 0.00 | 4.32 | 0.08 | 0.35 | 0.00 |
| China | 2.93 | 6.20 | 1.07 | 0.52 | 4.39 | 0.15 | 0.29 | 0.03 |
| Hong Kong | 0.26 | 0.00 | 0.00 | 0.00 | 2.37 | 0.00 | 0.00 | 0.00 |
| Japan | 6.26 | 0.51 | 0.00 | 0.00 | 1.88 | 0.00 | 0.00 | 0.00 |
| Korea | 2.31 | 0.00 | 0.00 | 0.00 | 6.64 | 0.00 | 0.00 | 0.00 |
| Taiwan | 2.14 | 0.00 | 0.00 | 0.00 | | | | |
| Indonesia | 0.76 | 6.88 | 2.30 | 7.13 | 4.66 | 0.66 | 2.54 | 1.61 |
| Malaysia | 0.81 | 0.00 | 1.07 | 3.22 | 9.70 | 0.00 | 2.32 | 1.43 |
| Philippines | 0.17 | 0.00 | 0.00 | 0.00 | 2.71 | 0.00 | 0.00 | 0.00 |
| Singapore | 1.02 | 0.00 | 0.00 | 0.00 | 16.42 | 0.00 | 0.00 | 0.00 |
| Thailand | 0.73 | 0.00 | 0.06 | 0.00 | 5.88 | 0.00 | 0.09 | 0.00 |
| Vietnam | 0.04 | 0.57 | 0.53 | 0.00 | 2.26 | 0.52 | 5.60 | 0.00 |
| Bangladesh | 0.02 | 0.00 | 0.00 | 0.00 | 0.61 | 0.00 | 0.00 | 0.00 |
| India | 0.66 | 0.01 | 0.03 | 0.00 | 2.10 | 0.00 | 0.02 | 0.00 |
| Sri Lanka | 0.04 | 0.00 | 0.00 | 0.00 | 2.99 | 0.00 | 0.00 | 0.00 |
| *Rest of South Asia* | 0.02 | 0.00 | 0.02 | 0.00 | 0.35 | 0.00 | 0.06 | 0.00 |
| Canada | 4.44 | 8.29 | 3.74 | 15.22 | 9.00 | 0.26 | 1.37 | 1.14 |
| US | 11.89 | 16.94 | 0.46 | 0.44 | 1.91 | 0.04 | 0.01 | 0.00 |
| Mexico | 1.10 | 0.00 | 4.58 | 0.05 | 3.62 | 0.00 | 2.72 | 0.01 |
| Central America and the Caribbean | 0.41 | 0.00 | 0.23 | 0.00 | 5.53 | 0.00 | 0.56 | 0.00 |
| Colombia | 0.16 | 4.27 | 1.01 | 0.00 | 2.14 | 0.90 | 2.47 | 0.00 |
| Peru | 0.23 | 0.00 | 0.11 | 0.00 | 4.49 | 0.00 | 0.38 | 0.00 |
| Venezuela | 0.32 | 0.90 | 4.90 | 0.00 | 4.86 | 0.22 | 13.51 | 0.00 |
| *Rest of Andean Pact* | 0.04 | 0.00 | 0.69 | 0.31 | 1.86 | 0.00 | 5.82 | 0.53 |
| Argentina | 0.29 | 0.00 | 0.93 | 0.11 | 1.13 | 0.00 | 0.66 | 0.02 |
| Brazil | 1.38 | 0.00 | 0.01 | 0.00 | 2.24 | 0.00 | 0.00 | 0.00 |
| Chile | 0.78 | 0.00 | 0.00 | 0.00 | 13.11 | 0.00 | 0.00 | 0.00 |
| Uruguay | 0.04 | 0.00 | 0.00 | 0.00 | 2.47 | 0.00 | 0.00 | 0.00 |
| *Rest of South America* | 0.05 | 0.00 | 0.00 | 0.00 | | | | |
| Austria | 1.34 | 0.00 | 0.00 | 0.00 | 8.72 | 0.00 | 0.00 | 0.00 |
| Belgium | 4.11 | 0.06 | 0.00 | 0.00 | 22.71 | 0.01 | 0.00 | 0.00 |
| Denmark | 0.70 | 0.00 | 0.39 | 0.00 | 5.48 | 0.00 | 0.55 | 0.00 |

| Country | % of total export, world prices | | | | Export relative to GDP | | | |
|---|---|---|---|---|---|---|---|---|
| | Energy-intensive goods | Coal | Oil | Gas | Energy-intensive goods | Coal | Oil | Gas |
| Finland | 1.34 | 0.00 | 0.00 | 0.00 | 14.69 | 0.00 | 0.00 | 0.00 |
| France | 6.40 | 0.10 | 0.07 | 0.27 | 5.96 | 0.00 | 0.01 | 0.01 |
| Germany | 11.32 | 0.26 | 0.19 | 0.27 | 7.02 | 0.00 | 0.02 | 0.01 |
| UK | 5.75 | 0.24 | 4.65 | 0.25 | 5.71 | 0.00 | 0.83 | 0.01 |
| Greece | 0.22 | 0.01 | 0.01 | 0.00 | | | | |
| Ireland | 1.20 | 0.00 | 0.00 | 0.00 | 22.95 | 0.00 | 0.00 | 0.00 |
| Italy | 4.56 | 0.00 | 0.02 | 0.00 | 5.17 | 0.00 | 0.00 | 0.00 |
| Luxembourg | 0.18 | 0.00 | 0.00 | 0.00 | | | | |
| Netherlands | 3.98 | 0.00 | 0.03 | 8.32 | 14.17 | 0.00 | 0.02 | 1.09 |
| Portugal | 0.38 | 0.00 | 0.00 | 0.00 | 4.93 | 0.00 | 0.00 | 0.00 |
| Spain | 2.13 | 0.03 | 0.00 | 0.00 | 5.23 | 0.00 | 0.00 | 0.00 |
| Sweden | 2.10 | 0.00 | 0.00 | 0.00 | 11.81 | 0.00 | 0.00 | 0.00 |
| Switzerland | 2.83 | 0.00 | 0.00 | 0.00 | 14.51 | 0.00 | 0.00 | 0.00 |
| *Rest of EFTA* | 0.79 | 0.03 | 7.94 | 7.97 | 6.15 | 0.00 | 11.17 | 2.29 |
| Hungary | 0.31 | 0.14 | 0.00 | 0.00 | 9.84 | 0.07 | 0.02 | 0.00 |
| Poland | 0.60 | 5.82 | 0.00 | 0.00 | 6.16 | 0.94 | 0.00 | 0.00 |
| *Rest of Central and Eastern Europe* | 1.51 | 2.12 | 0.01 | 0.00 | 14.80 | 0.32 | 0.01 | 0.00 |
| Former Soviet Union | 3.05 | 6.65 | 7.95 | 39.58 | 6.63 | 0.23 | 3.13 | 3.18 |
| Turkey | 0.48 | 0.00 | 0.00 | 0.00 | 3.20 | 0.00 | 0.00 | 0.00 |
| *Rest of Middle East* | 1.15 | 0.00 | 40.89 | 2.17 | 2.82 | 0.00 | 18.11 | 0.20 |
| Morocco, Western Sahara | 0.15 | 0.00 | 0.00 | 0.00 | 5.45 | 0.00 | 0.00 | 0.00 |
| *Rest of North Africa* | 0.21 | 0.00 | 5.38 | 10.62 | 1.56 | 0.00 | 7.27 | 2.93 |
| Botswana | 0.17 | 0.00 | 0.01 | 0.00 | 45.76 | 0.00 | 0.59 | 0.00 |
| *Rest of South African Customs Union* | 1.33 | 11.58 | 0.01 | 0.00 | 12.23 | 1.67 | 0.02 | 0.00 |
| Malawi | 0.00 | 0.00 | 0.01 | 0.00 | 0.18 | 0.00 | 0.46 | 0.00 |
| Mozambique | 0.00 | 0.00 | 0.00 | 0.00 | 0.60 | 0.00 | 0.00 | 0.00 |
| Tanzania | 0.00 | 0.00 | 0.00 | 0.00 | 0.52 | 0.00 | 0.00 | 0.00 |
| Zambia | 0.05 | 0.00 | 0.00 | 0.00 | 15.28 | 0.00 | 0.00 | 0.00 |
| Zimbawe | 0.05 | 0.02 | 0.00 | 0.00 | 8.10 | 0.06 | 0.00 | 0.00 |
| *Rest of Southern Africa* | 0.00 | 0.00 | 1.82 | 0.00 | 0.36 | 0.00 | 30.90 | 0.00 |
| Uganda | 0.00 | 0.00 | 0.01 | 0.00 | 0.37 | 0.00 | 0.48 | 0.00 |

| Country | % of total export, world prices | | | | Export relative to GDP | | | |
|---|---|---|---|---|---|---|---|---|
| | **Energy-intensive goods** | **Coal** | **Oil** | **Gas** | **Energy-intensive goods** | **Coal** | **Oil** | **Gas** |
| *Rest of Sub-Saharan Africa* | 0.19 | 0.08 | 7.42 | 0.00 | 1.55 | 0.01 | 10.94 | 0.00 |
| *Rest of World* | 0.42 | 0.00 | 0.72 | 1.58 | 1.95 | 0.00 | 0.61 | 0.27 |
| **Total** | **100.00** | **100.00** | **100.00** | **100.00** | | | | |

*Source*: Center for Global Trade Analysis (2001) (GTAP5)

*Note*: Blank cells indicate that no records are available.

# Part III

## External Enforcement – Parties and Non-parties

Chapter 6

# The Pros and Cons of External Enforcement[1]

Jon Hovi

## Purpose and plan

A number of scholars have pointed out that the level of compliance with international agreements in general, and international environmental agreements in particular, is largely quite good (Henkin, 1979; Brown Weiss and Jacobsson, 1999; Chayes and Chayes, 1993; 1995; Jacobsson and Brown Weiss, 1998). There are a number of potential explanations for this. According to one major view, states simply do not make decisions about compliance on the basis of a calculation of advantage. Instead, compliance is largely determined by factors such as standard operating procedures, internalized identities, norms of appropriate behaviour, domestic linkages and the perceived legitimacy of the relevant rules (Alter, 1998; Chayes and Chayes, 1993; 1995; Checkel, 2001; Franck, 1990; Mattli and Slaughter, 1998).

By contrast, a second view is that many international agreements do not require the parties to do much more than they would have done anyway, and states tend to adopt only regulations that are expected to achieve high levels of compliance. In short, states comply because there is little to be gained by non-compliance (Downs et al, 1996).

A third possible explanation relates international compliance to external enforcement. Consider the following statement by Canada's top negotiator at the 2001 climate meeting in Marrakesh: 'The matter [of a legally binding compliance mechanism for the Kyoto Protocol] is largely symbolic. Even if countries don't face sanctions, countries that signed on to the accord but did not abide by its rules would face pressure from other signatories'.[2] This statement suggests that the Canadian government was expecting the enforcement mechanism agreed upon in Marrakesh to have only marginal influence on compliance with the Kyoto Protocol, because in any case non-compliant behaviour would likely trigger external responses.

The purpose of this chapter is to review the strengths and weaknesses of external enforcement, and to discuss whether external enforcement has – or can have – distinct advantages over internal compliance mechanisms. Basically, I ask two questions. First, to what extent can it be useful to let external means of enforcement supplement an international agreement's internal compliance mechanism? Second, are there good reasons to restrict the use of external enforcement? Throughout, the argument is illustrated with examples from the Marrakesh Accords and other international agreements.

Although the possible impact of external enforcement has received surprisingly little attention in the compliance literature, at least three strands of relevant work may be identified. First, within the international law literature a number of authors have focused on countermeasures (e.g. Crawford, 2000; Kelly, 1998; O'Connell, 1994). For obvious reasons, however, this literature is mainly concerned with legal issues (such as the conditions under which countermeasures are permitted by international law), not the effectiveness of countermeasures in obtaining compliance. Also, there is only a partial overlap between the concept of countermeasures and that of external enforcement (see below). A second tradition analyses so-called unilateral sanctions (e.g. Chayes and Chayes, 1995, Chapter 4). While these authors focus at least partly on effectiveness, again the overlap with external enforcement is less than perfect. Finally, some work – starting with Schelling (1960; 1966) – either analyses how specific types of external enforcement can lend credibility to international agreements (e.g. Yarbrough and Yarbrough, 1986), or reviews a broader set of external enforcement techniques (e.g. Hovi, 1998, Chapter 7). However, this literature does not explicitly address the relative merits of external and internal enforcement.

This chapter proceeds as follows. The next section reminds the reader of what an international compliance system should ideally achieve. This is followed by a definition of the concept of external enforcement. The external–internal distinction is then related to two similar but not identical distinctions – namely collective versus unilateral enforcement, and centralized versus decentralized enforcement. This discussion provides a basis for the remaining parts of the chapter, which offer a treatment on the advantages and disadvantages of external enforcement.

## The purpose of a compliance system

In order to discuss in a meaningful way the advantages and disadvantages of external enforcement, we need to have a reasonably clear picture of what we would like a compliance system to achieve. The discussion in this chapter is based on the presumption that an ideal compliance system ought to have four characteristics.

First, it should deter intentional violations. In some cases a non-compliant party can achieve significant benefits, given that other parties continue to honour their obligations. It would thus be naïve to exclude the possibility that at least some parties might seize a tempting opportunity to cheat if this can be done without significant risk of getting punished. An ideal compliance system should therefore signal unambiguously that whatever gains can be obtained through deliberate non-compliance will be offset by subsequent penalties. To achieve this goal, it needs to have at its disposal effective means of verification as well as a set of credible and relatively severe response options.

Yet it is unlikely that even the most sophisticated enforcement mechanism would be able to deter *all* cases of non-compliance. For example, deterrence may be ineffective when non-compliance stems from a deficit in domestic regulatory capacity (Chayes and Chayes, 1995, pp13–15).[3] In cases of this sort, 'soft' responses might be more helpful than 'hard' consequences. An ideal compliance system should therefore have both types of measures at its disposal. In particular, it should be able to offer assistance and encouragement to parties that are willing but incapable of compliance.

Third, we would like a compliance system to restore damage. Ideally, it should arrange for compensation for victims of other countries' non-compliance. One reason why this is desirable is that a proper compensation scheme is likely to make an agreement renegotiation-proof,[4] thereby promoting effective deterrence. In addition, if a regime seeks to compensate for damages, it is more likely to be perceived as basically fair and legitimate.

Finally, an ideal compliance system should be able to live up to the requirements of due process. This means that decisions about enforcement and non-compliance should be based on fair procedures, like cases should be treated alike, and all parties should be equal before the court. Furthermore, the accused party should have a right to have its views heard, the burden of proof should normally rest with the party accusing another party of non-compliance, penalties should be imposed only after guilt has been proven beyond reasonable doubt, and there should be a possibility for appeal.

In summary, an ideal compliance system is one that deters deliberate non-compliance, assists parties that involuntarily fail to comply, compensates damage and achieves all this by way of due process. References in the following discussion to the advantages (pros) and disadvantages (cons) of external enforcement refer to whether external enforcement is likely to promote or hinder the accomplishment of any of these goals. Note that these goals are interrelated, in the sense that the accomplishment of one goal might have an impact on the likelihood of achieving another. For example, as previously mentioned, an arrangement that requires a non-compliant party to provide compensation to other parties can make an agreement renegotiation-proof, thereby contributing to the effective deterrence of non-compliance. Similarly, to the extent that regime assistance increases a party's capacity for compliance, it might also make deterrence more effective. Conversely, strong deterrents provide an incentive for a party to develop and increase its capacity for compliance. Finally, there may sometimes be a trade-off between effective deterrence on the one hand and due process on the other, as we shall see in later parts of the chapter.

## The nature of external enforcement

For the purposes of this discussion, enforcement is said to be external (i.e., external to the relevant agreement) if it involves either the resort to an enforcement mechanism that is not addressed by the relevant agreement, or a third party (such as a state or an organization) which is not itself a signatory (Hovi, 1998, p78).[5] In the case of the Kyoto Protocol, external enforcement might encompass the imposition of costs by Protocol parties on a non-compliant party using means not regulated in the protocol. It can also consist of non-governmental organizations (NGOs) or other non-signatories taking steps to punish countries that fail to fulfil their obligations.

Probably the most oft-cited type of external enforcement in relation to international environmental agreements is economic sanctions. However, if country A wants to punish country B's non-compliance with a given agreement externally, it has an abundance of other options as well. Here are a few examples of actions that country A might take:

- postpone or cancel a prior understanding to start bilateral negotiations over some issue of particular interest to country B;
- harden its position in an already ongoing negotiation with country B;
- decide against a company from country B that is competing for a major public contract in country A;
- introduce legislation or taxes that would be damaging to companies from country B;
- vote against country B's candidate for a major office in an international organization;
- withhold financial aid that country B would otherwise have received; and/or
- nullify preferential treatment that country B would otherwise have received.

The opposite of external enforcement is internal enforcement. Under many international agreements, the most severe internal sanction imaginable is termination of the relevant agreement. More typically, internal enforcement tends to involve the suspension of particular privileges, such as the suspension of the right to vote in a decision-making body. In the case of the climate regime, internal enforcement consists of responses to non-compliance on issues that are regulated by the Kyoto Protocol and that do not require involvement by a third party. Internal responses encompassed by the Marrakesh Accords include reductions of national allowances in subsequent commitment periods, and strictures on the right to participate in emissions trading.

The distinction between external and internal enforcement correlates, but is not congruent, with two other important distinctions. The first is the distinction between collective and unilateral enforcement, which has to do with the way in which the decision of whether to punish a non-compliant party is reached. If this decision is made through a procedure that has been pre-established by the parties through a collective decision, this is considered collective enforcement.[6] If the decision is not reached collectively, enforcement is said to be unilateral. Hence, this notion of unilateral enforcement corresponds roughly with Chayes and Chayes's notion of 'unilateral sanctions', where 'sanctions are not authorized by collective decision in accordance with agreed procedures' (Chayes and Chayes, 1995, p334, footnote 2).[7]

The second distinction is the one between centralized and decentralized enforcement, which concerns the source of the final decision to punish a party that has been found to be in non-compliance. If this decision is made by an international institution, such as the Compliance Committee or the Meeting of the Parties, this is said to be centralized enforcement. If, on the other hand, the decision is made individually by each party, enforcement is decentralized.[8]

An objection might be raised to this threefold set of distinctions on the grounds that external, unilateral and decentralized enforcement are likely to appear in combination, just as internal, collective and centralized enforcement might be expected to come as a package. For example, the treatment offered by Chayes and Chayes (1995, Chapter 4) on unilateral sanctions focuses almost entirely on sanctions that are not only unilateral, but also external and decentralized. Still, the point is easily overstated. It is not difficult to give examples of international compliance mechanisms that belong to opposite ends of the three distinctions. For example, the World Trade Organization (WTO) has a collectively adopted, well specified and indeed very detailed procedure for dealing with non-compliance and dispute settlement. Hence, in this organization enforcement is collective. Moreover, WTO rules restrict the spectrum of legitimate consequences to the suspension of the application to the non-compliant party of concessions or other obligations 'under the covered agreements'.[9] Accordingly, the organization relies on internal enforcement. However, the WTO's Dispute Settlement Body has no response option of its own. All it can do is to authorize the complaining party to suspend a concession or obligation in relation to the non-compliant country. Whether this punishment will actually be implemented is then up to that complaining party itself. Thus, the WTO mechanism for dispute settlement relies on internal, collective and decentralized enforcement.

A second example may be drawn from the regulation of international fisheries. In 1995 and 1996 the International Commission for the Conservation of Atlantic Tunas (ICCAT) adopted procedures for dealing with non-compliance. These procedures empowered the Commission – as a last resort – to recommend that contracting parties impose trade sanctions on a contracting (or non-contracting) party that has been found to be in non-compliance with ICCAT regulations (Carr, 1997). As in the case of the WTO, the final decision of whether or not to implement sanctions rests with the parties themselves; however, an important difference is that ICCAT sanctions may be unrelated to the issues regulated by the organization, meaning that in this case external, collective and decentralized enforcement is involved.

A final illustration stems from the enforcement of arms control treaties. In the 1980s, the US Assistant Secretary of Defense Richard Perle proposed that the Reagan administration adopt a systematic tit-for-tat strategy, whereby the US would promptly retaliate in kind every time the Soviet Union violated an arms control treaty between the two superpowers (Chayes and Chayes, 1995, p99). The proposal was never adopted (for good reasons), but the suggested approach was one of internal, unilateral and decentralized enforcement.

As these illustrations demonstrate, the three distinctions of internal/external, collective/unilateral and centralized/decentralized are not congruent but somewhat correlated. Because of the correlation, the advantages and disadvantages of unilateral or decentralized enforcement will sometimes also extend to external enforcement, as the following discussion will show.

## The pros of external enforcement

Can external means of enforcement be a useful supplement to the internal compliance mechanism of an international agreement, such as the Enforcement Branch of the Marrakesh Accords? This section argues that external means of enforcement can add significantly to the overall deterrent effect of a compliance regime. In some cases this is because external threats of punishment are likely to imply more severe consequences for the non-compliant country. In others it is because external threats carry more credibility than internal consequences. These two types of effect shall be discussed in turn.[10]

### *Severity of consequences*

Joyner (1998, p280) observes that an initial step in achieving compliance with international regulations is to establish a management regime that is acceptable to all parties. In other words, internal enforcement systems are generally products of international bargaining. As a result, they are likely to be subject to the so-called 'law of the least ambitious programme' (Underdal, 1980; 1998). This law, which is arguably of particular relevance for constitutional bargaining (e.g., bargaining about the formation and design of an international regime), says that in international bargaining the final word typically rests with the least enthusiastic party. It follows that we are well advised to expect the parties to international agreements to settle for relatively unambitious enforcement schemes. Anecdotal evidence seems to support this hypothesis. Relatively few international regimes and organizations have hard enforcement instruments of their own. Instead, internal enforcement is often limited to soft measures such as reporting, persuasion and shaming (Chayes and Chayes, 1995; Yoshida, 1999). To the extent that internal schemes for enforcement prove insufficient, external enforcement may be able to offer potentially more severe options. Alternatively, external reactions may be seen as a useful supplement to internal mechanisms for enforcement, thereby adding to the aggregate deterrent effect.

Effective deterrence requires that the expected response to non-compliance is more than proportional (Heister et al, 1997; Hovi and Areklett, 2004). Since future payoffs are likely to be discounted, the deterrent effect of a given punishment is not only a function of the size of the penalty, but also of the expected time interval between a violation and the implementation of the punishment. For any given response, the condition of more-than-proportional punishment is more likely to be achieved the more promptly the punishment is carried out. It is well known that procedures for internal enforcement can be very time-consuming. For example, prior to 1995 the GATT panel procedure was often criticized for its tardiness, taking an average of 18 months to settle a case (Hudec, 1990, p194). When the procedure underwent a fundamental revision in connection with the establishment of the WTO in 1995, this issue was a main concern. Under the new procedure, the timespan from the appointment of a panel to the rendering of a final decision should normally not exceed nine months (12 months if there is an appeal).[11] While obviously the total interval from violation to punishment remains longer, the new procedure is no doubt faster than the one which preceded it. This revision is probably one important reason why the new panel procedure has become vastly more popular than its predecessor.[12]

The combination of slow regime procedures with relatively modest consequences can sometimes be deliberately exploited. In the late 1990s, Norway introduced a policy of reserving certain positions at Norwegian universities for women, with a view to enlarging the proportion of female to male faculty members. In August 2000, complaints about this policy were submitted to the EFTA (European Free Trade Association) Surveillance Authority (ESA),[13] which in November 2001 found the policy to be in violation of the EEA (European Economic Area) principle of free competition in an open labour market. In March 2002, the ESA decided to bring the case before the EFTA court. Meanwhile, Norway continued to reserve positions for women, officially challenging the ESA's view of the practice as being inconsistent with EEA rules. Needless to say, in this case time worked in favour of the Norwegian government, in the sense that the longer the practice continued, the closer the government came to achieving its goal of a more gender-balanced faculty. Even if we assume that the Norwegian government realized that eventually it would have to give up the practice, it was certainly significant that this happened later rather than sooner. On the other hand, it is an open question whether the Norwegian government would have continued its practice for as long as it actually did, had the consequences of a conviction in the EFTA court been more severe.

Thus, to the extent that internal enforcement relies on collectively adopted procedures and due process, it tends to be time-consuming. This can, in turn, be a potential problem for deterrence. By contrast, external enforcement is often unilateral and therefore more easily compatible with expediency. It follows that external means of enforcement are likely to be relatively popular in cases where internal procedures are slow. On the other hand, if internal procedures are revised to become faster, more countries are likely to choose internal over external enforcement, as illustrated by the General Agreement on Tariffs and Trade (GATT)/WTO example.

Finally, internal schemes for enforcement are not always self-supporting. At one end, we have cases where the deterrent effect of internal measures is explicitly linked to external elements. For example, the Seabed Disputes Chamber of the International Tribunal for the Law of the Sea does not have any means of enforcement of its own. However, according to the statutes of the tribunal, the Chamber's decisions 'shall be enforceable in the territories of the State Parties in the same manner as judgments or orders of the highest court of the State Party in whose territory the enforcement is sought' (Article 39).

At the other end, we have cases where a regime or an organization is entitled to revoke some kind of privilege from a party that has been deemed to be in non-compliance. For example, the UN Charter says that 'a member of the organization which is in arrears in the payment of its financial contributions to the Organization shall have no vote in the General Assembly if the amount of its arrears equals or exceeds the amount of the contributions due from it for the preceding two full years' (Article 19). Similarly, a member of the UN against which preventive or enforcement action has been taken by the Security Council 'may be suspended from the exercise of the rights and privileges of membership by the General Assembly upon the recommendation of the Security Council' (Article 5). Finally, a member that has persistently violated the principles of the UN Charter 'may be expelled from the Organization by the General Assembly upon the recommendation of the Security Council' (Article 6). Given that membership is voluntary, we may safely assume that members see it as desirable to retain privileges of this kind. To the extent that a threat to revoke privileges is credible,[14] it therefore provides a clear-cut incentive for compliance.[15] And, unlike the mechanisms of the United Nations Convention on the Law of the Sea (UNCLOS), this incentive does not depend on external support of any kind.

At first sight, the enforcement mechanism found in the Marrakesh Accords appears to fall in the second category, i.e., the one of self-supporting internal enforcement. After all, revoking the right to participate in emissions trading, or reducing the assigned amount of emissions in future periods, looks much like suspending the right to vote, inasmuch as both types of penalty imply that a privilege is revoked (in part or in full). On closer inspection, however, there is an important difference. Subtraction of excess emissions from the assigned amount for the subsequent commitment period is costly only to the extent that a party sticks to its commitment to comply with the assigned amounts of future periods. If it should simply decide to ignore these obligations, there is little that the climate regime itself would be able to do. Similarly, if a country has its right to participate in emissions trading suspended, and responds by withdrawing from the Kyoto Protocol, the regime would be largely powerless.[16] To be effective, therefore, both response options depend on support from external enforcement mechanisms. In principle, a party that has been punished by the Enforcement Branch could nullify the punishment by ignoring future obligations or withdrawing from the regime. In practice, however, external pressure from other countries may induce the party to adhere to its future commitments, thereby making the internal punishment imposed by the Enforcement Branch effective.

## *Credibility*

There are exceptions to the rule that international regimes typically lack powerful means of enforcement. In a number of these regimes, however, the enforcement schemes are basically paper tigers that are rarely if ever actually used. Of course, the fact that a mechanism for enforcement is rarely used is not by itself proof that it is toothless. In fact, a mechanism that provides effective deterrence is likely to be used only exceptionally. However, effective deterrence is – at best – only part of the explanation why international enforcement schemes are not often used. The truth is that in a number of cases these schemes are not used *despite* the fact that they often fail to deter.

A possible explanation is that these internal enforcement mechanisms are riddled with credibility problems. One important reason for this lack of credibility is that decisions about internal enforcement are often made collectively by a political agency. For example, in the Montreal Protocol, decisions as to what kind of recommendations should be made, and which measures should be taken, are left 'largely to the broad discretion of the Implementation Committee and the Meeting of the Parties' (Yoshida, 1999, p140). Needless to say, governmental representatives cannot necessarily be expected to act impartially, or to always let professional standards dominate political motives. Depending on the decision-making procedures of the relevant body, a well-founded proposal to punish a state that fails to honour its obligations may therefore be blocked by that state itself, by its allies, or by countries that are simply afraid to jeopardize their relations with the non-compliant state. Thus, even in the event that a majority of the Member States wish to impose a penalty, this could well prove impossible to achieve through internal procedures.

In cases of this sort, external enforcement may be a more credible deterrent. There are a number of reasons for this. First, external enforcement is often also unilateral, meaning that the decision to impose some kind of punishment does not depend on the consent of other countries. In some cases a government may even be required by law to consider unilateral punishment. An example is the US's Marine Mammal Protection Act (MMPA), which forbids the import into the US of fish caught by fishermen from countries that do not protect marine mammals. Similarly, the Pelly Amendment to the Fishermen's Protective Act of 1967 imposes US sanctions against countries that diminish the effectiveness of the conservation programme of the International Whaling Commission (IWC). According to the Pelly Amendment, the President is required to explain and justify any departure from such action.[17] No doubt legislation of this kind serves to make a threat of external enforcement more credible than it would otherwise have been.[18]

A second reason why external enforcement may enhance credibility is that it sometimes reverses the burden of proof. Suppose, for example, that a country unilaterally introduces economic sanctions to punish a party that has failed to honour its obligations under the Kyoto Protocol. While potentially this might constitute a violation of WTO regulations, it would be up to the country being sanctioned to bring the matter before a WTO panel to obtain a ruling that a violation of a WTO agreement has taken place. Since the main rule in WTO is that quantitative restrictions on trade (such as a boycott) are illegal, the country imposing sanctions would have to establish that one of the exception clauses applies.[19] As we have already seen, however, the panel procedure takes time. In the meantime the country being sanctioned (i.e., the country that initially violated the Kyoto Protocol) suffers. By contrast, if a country seeks to punish a violation by internal means, it must proceed through procedures that at worst end up in nothing, and at best result in a penalty after considerable delay.

Finally, even if it is clear that (a certain type of) external enforcement is illegal, a ban on this form of enforcement may itself be difficult to enforce. In particular, it may be difficult to verify that a given measure has been undertaken *for the purpose of* punishing a violation. For example, in March 2002 the US announced that it would impose a prohibitive import tax on steel to protect its domestic steel industry 'in a transition period'. A few weeks later, the EU Commission launched a proposal involving new taxes and limitations on landing rights for non-EU airlines that were exploiting government subsidies to dump prices on flights to and from the EU area. This proposal was likely to hurt several US airlines, which had received massive subsidies after the airborne terrorist attacks in the US on September 11, 2001. Thus, media reports immediately suggested a link between the EU measure and the US steel tax. US officials declared that any complaints should be launched through WTO channels. However, EU spokesmen officially rejected a link between the measures, stating that the anti-subsidy proposal had been underway for some time before the imposition of the steel tax.

## The cons of external enforcement

The title of this chapter, as well as the subtitle of this section, suggests that external enforcement is something that the parties to an international agreement can potentially rid themselves of, if they so choose. Indeed, it might be argued that the provision of an internal enforcement mechanism can also be understood as an understanding to abstain from external enforcement.

While this view might have some intuitive appeal, in practice it is extremely difficult to eliminate the option of external enforcement. As illustrated by the case of the US steel tax, it may be almost impossible to verify that a given external measure, although undisputedly detrimental to a party that has been accused of non-compliance, has been imposed for that particular purpose. It is not hard to envision cases where the non-compliant party knows exactly what is going on, but finds itself unable to prove it in front of an international dispute settlement body.[20]

What *is* feasible, however, is to impose regulations with a view to making external enforcement more difficult or costly than it would be in the absence of such inhibitory regulation. Two types of regulation are conceivable. First, institutions or agreements outside the climate regime might include provisions that restrict the parties' freedom to enforce the Kyoto Protocol externally. Second, strictures on the use of external enforcement might be provided by the climate regime itself.

### Outside restrictions

Why should treaty A impose restrictions on the ways in which treaty B may be enforced? There are at least two possible answers. The first is that certain types of action are simply deemed unacceptable. For example, the UN Charter places a general ban on the use of armed force except in self-defence. According to the International Court of Justice, this ban means that states may use force only in response to an armed attack. Thus, even in the (hypothetical) case where a state is faced with an imminent and potentially fatal threat resulting from a breach of an environmental agreement (such as the Kyoto Protocol), the state would not be entitled to use armed force (O'Connell, 1994, p4).

The second answer is that the enforcement of treaty B might be detrimental to the fulfilment of treaty A. For example, one of the fundamental principles of the GATT is the ban on discriminatory restrictions on trade. According to this principle, all members of the GATT are entitled to most favoured treatment. The use of trade sanctions against a country that has failed to honour its obligations under the Kyoto Protocol would potentially violate this rule.[21] Thus, the GATT bans external enforcement via trade (unless one of the exception clauses applies).

Of course, we cannot automatically infer from the fact that treaty A places a ban on external enforcement of treaty B that such enforcement will not take place. A major focus of this book is to investigate the conditions under which the parties are likely to comply with the Kyoto Protocol. It goes without saying that for such an investigation to be of interest, it must be assumed that perfect compliance with the Kyoto Protocol cannot be taken for granted. It would then be a strange thing indeed to take it for granted that *other treaties* always obtain perfect compliance. This said, it is extremely unlikely that anyone would even consider the use of military force to enforce the Kyoto Protocol. It is less obvious, however, that the possibility of using trade sanctions will not be seen as an option either.

### Inside restrictions

An important reason why a treaty might seek to restrict the use of external enforcement is that such enforcement is considered to be detrimental to the regime's legitimacy. As Chayes and Chayes remind us, procedures for international enforcement ought to satisfy the same standards that apply to other law enforcement activities: 'The most fundamental of these standards are that like should be treated alike, that the crucial determinations should be made by basically fair procedures, and that all actors should be equal before the court' (Chayes and Chayes, 1995, p106). The Chayeses claim that unilateral sanctions are defective on all these counts. Their argument seems to be equally relevant for most cases of external sanctions. First, the legal issue is only one of many considerations for political leaders. Thus, legal considerations will typically be balanced against state interests, which are likely to vary from case to case and from country to country. It is therefore unlikely that unilateral sanctions will generally treat equivalent violations equally. Second, the decision to impose unilateral sanctions is made by the sanctioning government. To act as a judge in one's own case is a fundamental violation of due process of law. Third, unilateral sanctions are disproportionally used by major powers – notably the US – against economically and politically weaker counterparts. Hence, their exercise is likely to seem 'less like disinterested enforcement of common norms and more like the exertion of power in the interest of the stronger' (Chayes and Chayes, 1995, p108). Similarly, it has been pointed out that countermeasures are far more available to wealthy countries than to poor ones:

> States holding assets to freeze or markets to close can do both as countermeasures. Poor states will often have few resources available for leverage. Even when resources are available, these states may need to be able to withstand counter-countermeasures to effectively use the countermeasure. Thus some see countermeasures as fundamentally unfair and therefore question their acceptability in a legal system. (O'Connell, 1994, p4)

We might add that external enforcement usually violates other criteria of due process as well. For example, there is usually no guarantee that the accused party has a right to have its case heard. Nor is it clear that punishment will be imposed only after guilt has been proven beyond reasonable doubt. And even if it is, the standards of reasonable doubt are likely to be determined unilaterally. Finally, due process usually requires that there is a possibility for appeal. This is not likely to exist in most cases of external enforcement.

A second potential motive to impose regulations on the use of external enforcement is that such enforcement is unlikely to provide fair compensation for damages. Granted, it is *possible* that agreement on compensation can be reached by external means. For example, WTO rules instruct the parties in a dispute to seek mutually acceptable terms of compensation, and only if the parties prove unable to reach agreement may the Dispute Settlement Body authorize retaliation. Similar rules are – in principle – conceivable within the climate regime, even though arrangements for compensation that are reached bilaterally will often reflect the parties' bargaining power more than impartial principles of fairness.

This said, however, it needs to be pointed out that in the case of the climate regime, *internal* arrangements for compensation are likely to see some serious problems as well. Compensation usually requires that it be possible to establish a causal link between acts of non-compliance and the injury suffered by others. While inferences of this kind are plausible in some instances,[22] in the case of climate change there are potentially serious difficulties. First, it is almost impossible to determine whether a given instance of extreme weather is a random event or the result of global warming. Second, it is extremely difficult to distinguish between global warming from anthropogenic sources and natural climate change. Finally, man-made global warming is not attributable to any one state's activity. On the contrary, most states have only a minimal role in global emissions of greenhouse gases (GHGs). Together, these three facts make it unlikely that it is possible to establish a sufficient link of causation between GHG emissions from activities by particular states or companies on the one hand and direct injury to individual states on the other.[23] It is therefore likely that an arrangement for compensation under the climate regime would have to take a more rudimentary form. For example, one might envision a system whereby a reduction in a non-compliant country's allowance is transferred to other parties in relation to their quotas.

While I believe that the above objections to external enforcement are basically valid, one should note that the norms of equal treatment, fair procedures and equality before the court are not always met by internal enforcement schemes either. For example, an objection to the GATT/WTO dispute settlement procedure is that it is generally difficult for weak states to implement sanctions against the US or other major powers, even if a prior consent from the Dispute Settlement Body has been granted (van Bael, 1988, pp71–2). Thus, much of Chayes and Chayes's criticism of unilateral sanctions seems to hold for collective enforcement as well, at least if the implementation of the punishment is decentralized. When the US refused to comply with the orders of the International Court of Justice (ICJ) in the Nicaragua Case in 1986, there was little Nicaragua could do to get the ICJ's orders enforced (O'Connell, 1994). That the willingness to enforce international law differs across countries can also be seen in some environmental regimes. While all governments that ratify MARPOL (International Convention for the Prevention of Pollution from Ships) Conventions are in principle responsible for enforcing the requirements of the conventions, statistics show that actual performance varies considerably. The flag state with the worst accident record suffers more than a hundred times more losses than the flag state having the best record. In particular, ships registered under flags of convenience are more likely to cause accidents and pollution, to be in poor condition, to have communication problems due to multilinguistic personnel, and to have inadequately trained and certified crews (Becker, 1998, p634).

Chayes and Chayes warn that the use of unilateral sanctions is likely to undermine a regime's legitimacy. This certainly seems like a plausible hypothesis, and again one that might be expected to be equally valid for many cases of external enforcement. Yet one might reasonably ask why external (or for that matter unilateral) enforcement should be more detrimental to regime legitimacy than the violations it seeks to address. To the extent that the risk of facing external sanctions serves as an effective deterrent, it prevents violations that would otherwise have taken place. Certainly, this should serve to strengthen the regime. The price to be paid is that whenever deterrence fails, external sanctions must be carried out in order to preserve credibility. This creates the effect that the Chayeses are so worried about. Nevertheless, we might reasonably assume that the extent to which an act of external enforcement tends to undermine a regime's legitimacy is likely to depend on whether the act itself corresponds with the dictates of international law.

External sanctions may be permitted under international law as 'countermeasures'. To be acceptable, a countermeasure must meet certain conditions. First, it has to be *necessary*, in the sense that prior requests that the target state take the necessary steps to achieve compliance have proven ineffective. Second, the countermeasure must be *proportionate* to the unlawful act that it seeks to address. If taken literally, so that an act of non-compliance that produces a gain of US$10 million for the non-compliant state, for example, can only result in a penalty equivalent to (at most) the same amount, this requirement would render any lawful countermeasure ineffective as a deterrent. However, the condition of proportionality is probably more reasonably interpreted as saying that a countermeasure cannot be *vastly* out of proportion to the unlawful act that it seeks to address. While it is straightforward to give examples that would clearly violate this norm, it is difficult to say exactly where the line should be drawn.[24] A third condition for the lawfulness of a countermeasure, referred to in passing by the ICJ in 1997, [25] is that its purpose 'must be to induce the wrongdoing state to comply with its obligations under international law, and that the measure must therefore be reversible' (cited in Crawford, 2000, pp3–4). This condition means that permanent measures cannot qualify as countermeasures.[26]

To the extent that the above conditions of necessity and proportionality are satisfied, external enforcement is consistent with the dictates of international law.[27] If the relevant regime itself is incapable of ensuring compliance, it is difficult to see why external enforcement should then necessarily weaken the legitimacy of the regime.

## Conclusion

External enforcement may enhance the deterrent effect of an international regime, but at the same time it may operate to undermine the regime's legitimacy. However, this trade-off is likely to depend on both the circumstances that lead to the imposition of sanctions, and the external sanctions' lawfulness. If external enforcement takes place within the boundaries set by international law, while at the same time adding significantly to the deterrent effect of a regime's internal consequences, it is likely to enhance compliance without necessarily being detrimental to the regime's legitimacy. Although the best option of all is likely to be an internal system of enforcement that is both effective and fair, a second-best might option may be external enforcement in accordance with the dictates of international law.

## Notes

1 This chapter has greatly benefited from comments received at two seminars held in Oslo in January and May 2002 as part of the preparation of this book.

2 WSJ.com, November 27, 2001, http://interactive.wsj.com/fr/emailthis/retrieve.cgi?id=ID-CO-20011103-000334.djml

3 Note, however, that this does not necessarily make deterrence irrelevant. The latter would be true only if we assume that the capacity for compliance is fixed, not something that can

be developed. If capacity building is possible, then strong deterrents are likely to induce a country to invest more in capacity building than it would otherwise have done. On the other hand, if the capacity for compliance is fixed, it would make encouragement and assistance just as irrelevant as deterrence. Granted, it is in principle conceivable that others may be able to offer know-how about capacity building that the country itself does not have; but then we should at least ask why it is not possible for the country to buy this know-how in the first place. Thus, it seems that only under rather special circumstances can we have a situation where capacity building is possible, so that encouragement and assistance have a role to play, while at the same time deterrence is irrelevant.

4 For a discussion of renegotiation-proofness in relation to the Kyoto compliance system, see Hovi and Areklett, 2004.

5 This definition raises the following question: what if an environmental treaty explicitly states that the parties are allowed to use economic sanctions to enforce it, or that disputes about compliance shall be referred to a third party for settlement? It seems that in such a circumstance, enforcement would still be said to be external, and such rules would simply authorize and formalize the use of external means of enforcement.

6 The term 'collective' rather than 'multilateral' is used, since procedures of this kind may also be established to enforce bilateral agreements.

7 Chayes and Chayes concede that the term 'unilateral' is not wholly satisfactory, because the measures they have in mind are often taken by a number of countries in concert.

8 Yet another related distinction is Keohane's (1986) distinction between 'specific' and 'diffuse' reciprocity, where the former refers to tit-for-tat retaliation within the same situation, while the latter refers to punishment at another time, in another transaction. To the extent that these concepts only refer to responses made by other states, not by international institutions, diffuse reciprocity is a subset of external enforcement.

9 WTO understanding on rules and procedures governing the settlement of disputes, Art 22.2.

10 Note that a particular factor that is an advantage in one respect may easily become a disadvantage in another. In other words, there may be a price to be paid for obtaining a certain desirable consequence. Therefore, the decision of whether to describe a given factor as an advantage or disadvantage may sometimes be somewhat arbitrary.

11 See the WTO understanding on rules and procedures governing the settlement of disputes, Art 20.

12 Another important reason has to do with the fact that, prior to 1995, decisions were made by consensus, meaning that a party could veto a panel report supporting the complaining party. By contrast, under the new procedure, the consensus rule has been turned on its head, so that now a consensus is required to *block* a panel report from being adopted.

13 The task of the ESA is to ensure that EEA rules are properly enacted and applied by the EFTA states.

14 Granted, it is far from obvious that the above mentioned provisions of the UN Charter are credible. For example, in May 1999 as many as 17 member countries owed the organization more than the amount due from them for the preceding two years, yet none of these countries had their right to vote suspended (Hovi and Underdal, 2000, p52).

15 However, it is an open question whether or not a threat to revoke these privileges constitutes a more than proportional punishment. Needless to say, this is likely to depend on the potential gains that each country can obtain by violating the relevant clauses.

16 In this respect, the Kyoto Protocol resembles most other international treaties.

17 Still, there are several cases where Norway, Canada and Iceland have been 'certified' to be in violation of the Pelly Amendment without sanctions being imposed.

18 Other examples include the 1979 Packwood-Magnuson Amendment to the Fishery Conservation and Management Act of 1976, which allows the US to reduce or suspend fishing privileges in US waters for nations acting contrary to IWC guidelines. However, in the

case of whaling, the threat of Packwood-Magnuson sanctions is no longer influential, since no foreign whaling nation currently fishes in US waters.

19 For discussions of these exception clauses, see Kelly, 1998 and Chapter 7 by Stokke in this volume.

20 Note also that it is usually easier to enforce a general ban on a particular activity (such as the use of import taxes) than it is to prohibit that activity only for a specific purpose (such as external enforcement of a particular treaty).

21 For a detailed discussion, see Chapter 7 by Stokke.

22 A case in point may be when an oil spill from a tanker pollutes a coastal area.

23 In this respect, the climate regime resembles the ozone regime. See Yoshida, 1999, p98.

24 For example, in the early 1990s the US cut off imports of tuna caught by Mexican fishermen on the grounds that the Mexican harvesting of tuna incidentally killed a large number of dolphins. While apparently accepting this boycott as proportional, O'Connell (1994, p4) suggests that in this case a general trade embargo would have been out of proportion.

25 The case concerned the Gabcikovo–Nagymaros project. Here, the court also added a clarification to the principle of proportionality by stating that 'the effects of a countermeasure must be commensurate with the injury suffered, taking account of the rights in question' (cited in Crawford, 2000, footnote 9).

26 However, this again recalls an important problem: that it may be difficult to establish the real motive behind an external sanction. For example, it may be extremely difficult in practice to establish whether the real motive behind an import tax is a legitimate intention to induce the target state to comply with the Kyoto Protocol, or simply a desire to protect a domestic industry against foreign competition.

27 Note that these conditions refer to enforcement based on otherwise unlawful actions (reprisals). Enforcement based on unfriendly but lawful actions (retortions) is always allowed by international law (O'Connell, 1994, p4). An example of a retortion is the withdrawal of diplomats.

## References

Alter, K. A. (1998) 'Who Are the "Masters of the Treaty"? European Governments and the European Court of Justice', *International Organization*, vol 52, 121–47

Bacon, B. L. (1999) 'Enforcement Mechanisms in International Wildlife Agreements and the United States: Wading Through the Murk', *Georgetown International Environmental Law Review*, vol 12, 331–63

Baylis, J. (2002) 'Arms Control and Disarmament' in Baylis, J., Wirtz, J., Cohen, E. and Gray, C. S. (eds) *Strategy in the Contemporary World*, Oxford University Press, Oxford, 183–207

Becker, R. (1998) 'MARPOL 73/78: An Overview in International Environmental Enforcement', *Georgetown International Environmental Law Review*, vol 10, 625–42

Brown Weiss, E. and Jacobsson, H. (1999) 'Getting Countries to Comply with International Agreements', *Environment*, vol 41, no 6, 16–20, 37–40

Carr, C. J. (1997) 'Recent Developments in Compliance and Enforcement for International Fisheries', *Ecology Law Quarterly*, vol 24, 847–60

Chayes, A. and Chayes, A. H. (1993) 'On Compliance', *International Organization*, vol 47, 175–205

Chayes, A. and Chayes, A. H. (1995) *The New Sovereignty: Compliance with International Regulatory Agreements*, Harvard University Press, Cambridge, MA

Checkel, J. (2001) 'Why Comply? Social Learning and European Identity Change', *International Organization*, vol 55, no 3, 553–88

Crawford, J. (2000) 'The Relationship between Sanctions and Countermeasures', www.law.cam.ac.uk/rcil/ILCSR/sanccms99(f).doc)

Downs, G. W., Rocke, D. M. and Barsoom, P. N. (1996) 'Is the Good News About Compliance Good News About Cooperation?', *International Organization*, vol 50, 379–406

Franck, T. M. (1990) *The Power of Legitimacy Among Nations*, Oxford University Press, New York, NY

Heister, J., Mohr, E., Stähler, F., Stoll, P.-T. and Wolfrum, R. (1997) 'Strategies to Enforce Compliance with an International $CO_2$ Treaty', *International Environmental Affairs*, vol 9, 22–53

Henkin, L. (1979) *How Nations Behave* (second edition), Columbia University Press, New York, NY

Hovi, J. (1998) *Games, Threats and Treaties: Understanding Commitments in International Relations*, Pinter, London

Hovi, J. and Areklett, I. (2004) 'Enforcing the Climate Regime: Game Theory and the Marrakesh Accords', *International Environmental Agreements: Politics, Law and Economics*, vol 4, 1–26

Hovi, J. and Underdal, A. (2000) *Internasjonalt Samarbeid og Internasjonal Organisasjon*, Universitetsforlaget, Oslo

Hudec, R. E. (1990) 'Dispute Settlement' in Schott, J. J. (ed) *Completing the Uruguay Round*, Institute of International Economics, Washington, DC, 180–204

Jacobsson, H. and Brown Weiss, E. (1998) *Engaging Countries: Strengthening Compliance With International Environmental Accords*, MIT Press, Cambridge, MA and London

Joyner, C. C. (1998) 'Compliance and Enforcement in New International Fisheries Law', *Temple International and Comparative Law Journal*, vol 3, 271–300

Kelly, A. J. (1998) 'The GATT Obstacle: International Trade as an Obstacle to Enforcement of Environmental Conservation on the High Seas', *Florida Journal of International Law*, vol 12, 153–66

Keohane, R. (1986) 'Reciprocity in International Relations', *International Organization*, vol 40, 1–27

Koh, H. H. (1997) 'Why Do Nations Obey International Law?', *Yale Law Journal*, vol 106, 2599–659

Mattli, W. and A. M. Slaughter (1998) 'Revisiting the European Court of Justice', *International Organization*, vol 52, 177–209

O'Connell, M. E. (1994) 'Using Trade to Enforce International Environmental Law: Implications for United States Law', *Indiana Journal of Global Legal Studies*, vol 1, no 2, http://ijgls.indiana.edu/archive/01/02/oconnell.shtml

Schelling, T. S. (1960) *The Strategy of Conflict*, Harvard University Press, Cambridge, MA

Schelling, T. S. (1966) *Arms and Influence*, Yale University Press, New Haven, CT

Underdal, A. (1980) *The Politics of International Fisheries Management: The Case of the North-East Atlantic*, Scandinavian University Press, Oslo

Underdal, A. (1998) 'Introduction' in Underdal, A. (ed) *The Politics of International Environmental Management*, Kluwer, Dordrecht, pp 1–12

van Bael, I. (1988) 'The GATT Dispute Settlement Procedure', *Journal of World Trade*, vol 22, no 4, 67–77

Yarbrough, B. V. and Yarbrough, R. M. (1986) 'Reciprocity, Bilateralism and Economic "Hostages": Self-enforcing Agreements in International Trade', *International Studies Quarterly*, vol 30, 7–21

Yoshida, O. (1999) 'Soft Enforcement of Treaties: The Montreal Protocol's Non-compliance Procedure and the Functions of Internal International Institutions', *Colorado Journal of International Environmental Law and Policy*, vol 10, 95–141

Chapter 7

# Trade Measures, WTO and Climate Compliance: The Interplay of International Regimes

Olav Schram Stokke

## Introduction

The purpose of this chapter is to examine how the potency of a particular external compliance mechanism – trade measures – is affected by the interplay between multilateral trade rules and the global climate regime.[1] Examples of such measures could be border taxes or levies imposed upon climate-related goods that originate in countries with more lenient climate regulations than those of the importing country. Such trade measures would be an external compliance mechanisms because there are currently no provisions in the climate regime that explicitly mandate or authorize restrictions on the cross-border flow of goods and services, either on non-parties to the Kyoto Protocol or on parties that fail to comply with their commitments. It is true that the Marrakesh Accords restrict trade in emissions allowances since only parties with emissions-reduction commitments may engage in such trade, and even they can have their eligibility suspended, or limited to purchase, if they are found to be in non-compliance.[2] However, such allowances are licences issued by a government in order to regulate certain activities within its territory, and the fact that they can be traded internationally does not make them into goods or services as understood by the World Trade Organization (WTO) (Werksman, 1999, p5; Charnovitz, 2003, p152).

The relevance of unilateral, uncoordinated trade measures for environmental purposes has been debated extensively for decades. The issue goes beyond compliance, as such measures can also be used to encourage non-parties to an international environmental regime to join it (Goldberg et al, 1997). The US in particular has been active in implementing unilateral trade measures justified by environmental goals.[3] In the late 1970s, the Pelly Amendment to the Fisherman Protection Act authorized the use of trade sanctions, in practice related to seafood products, against states which undermined international conservation agreements. This feature has been prominent in the overall compliance system of the international whaling regime (Friedheim, 2001). More recently, several US trade measures to discourage foreign bycatchintensive harvesting methods in tuna and shrimp fisheries have been challenged under the WTO rules (Joyner and Tyler, 2000).

Multilaterally coordinated trade measures, mandated or authorized by international agreements, are less vulnerable to challenge under international trade rules, especially if the target of sanctions is also a party to the agreement (Oberthür and Ott, 1999, p280). Any discrimination would have to be justified in explicit WTO exceptions to the general rules.[4] Around 20 out of an estimated 200 multilateral environmental agreements provide for trade-related compliance measures today.[5] While environmentalists tend to consider provisions for green trade measures as valuable reinforcements of the generally weak institutions of international environmental governance, such provisions are highly controversial. Many developing countries point out that as a policy instrument, trade measures are asymmetrically available because their effectiveness depends heavily on the size and diversity of one's home market, and thus tends to favour states which are already powerful (Bhagwati, 2002; Charnovitz, 2002a, p428). Conversely, there is concern that provisions for environmental trade measures may be abused for protectionist purposes and gradually undermine the liberalist norms of the global trade regime (Neumayer, 2001, pp150–1).

The relationship between the global trade regime and the emerging climate regime is multifaceted and only parts of it are pertinent to the discussion of trade measures as compliance instruments.[6] This chapter will address restrictions introduced by a state or a group of states (enforcer) on imports of climate-related goods from another state (target) in order to facilitate or induce more climate-friendly policies on the part of the enforcer, target or third parties.[7] Such climate-friendly policies could include participation in the Kyoto Protocol or improved compliance with commitments under the Protocol. The participation issue is likely to become relevant with the beginning of the first commitment period (2008–12), whereas non-compliance can only be ascertained, and thus become an issue relevant for enforcement, several years after the expiry of that period. The inclusiveness with regard to whose policy adaptation is intended reflects the fact that, unlike the case of trans-boundary air pollution, it is the aggregate greenhouse gas (GHG) emissions that matter and not their geographic patterns. Even in cases where a target of trade measures refuses to change its ways, the costs incurred by those measures may impress third parties and thus be relevant to the problem of global warming.[8]

## Trade measures and climate behaviour

Trade measures can influence the attainment of environmental objectives in a number of ways. In cases such as the listing of certain species under the Convention on International Trade in Endangered Species of Wild Fauna and Flora (CITES), checking international trade flows is an objective in itself because the existence of a global market generates incentives for the activity which causes environmental damage, notably game poaching. Similarly, trade controls introduced under the Montreal Protocol on ozone are designed to help parties fulfil their obligations to phase out the domestic production and use of ozone-depleting substances, which is generally complicated by cross-boundary flows of such substances.[9]

In the climate context, trade measures are likely to be justified not by the need to diminish a market, but to offset any competitive advantage held by producers in countries that have not implemented costly mitigation measures. At least in some sectors, differences among states in climate policies may affect their relative costs of production (Fox, 1996, p2513–16) and it would be surprising if this does not generate demands for a more level playing field. Sectors that rely heavily on export markets may call for more lenient implementation of domestic climate measures or require that the government pick up a greater part of the climate bill, for instance by subsidizing investments in green technology and equipment. Sectors that face heavy competition from imports are likely to request import restrictions to ensure that any advantage foreign producers gain from less-stringent domestic climate measures are removed at the border. This could happen even if the states in question are both Kyoto compliers but meet their commitments by a different mix of policies and measures (Chambers, 2001).

In cases where carbon taxation is part of the domestic climate policy, offsetting measures may take the form of border taxes.[10] In other cases, a countervailing levy could be placed on carbon-intensive products that originate in economies with more lenient climate restrictions; the EU has reportedly informally considered the possibility of introducing such a levy (Jacob, 2001). Referring to environmental agreements in general and the Kyoto Protocol in particular, a policy paper issued by the EU Commission stated that

> if a country imposed a trade measure for environmental purposes on another WTO Member which had not signed the MEA... [c]ould the country affected use WTO rules to overrule the trade measure? The EU wants WTO Members to agree that this should not happen. (EU Commission, DG Trade, 2001)

A third possible offsetting measure is energy efficiency standards. So far, such standards have only been used with regard to product characteristics – for instance, automotive fuel economy or the energy efficiency of various household appliances – but they could also be devised to differentiate products on the basis of the energy used in the production process (Brack et al, 2000, p43). An energy efficiency standard could be implemented to keep products that fail to meet the requirements outside the domestic marketplace, and border taxes and levies would increase the price of the imported product.

The causal processes that connect these instruments with more climate-friendly policies would be essentially the same. The enforcer would worry less about loss of competitiveness when implementing its climate commitments (Charnovitz, 2002b, p62). If the enforcer is a major market or persuades others to introduce similar trade measures, the target would no longer have incentives to stay outside of a climate regime (external free-riding) or ignore its obligations under it (internal free-riding). Third parties would seriously consider introducing environmental measures to avoid being targeted for sanctions in the future. Taken together, this would reduce the risk of the leakage, through effects on commodity prices, of emissions-intensive production from those who assume and implement costly mitigation measures to those who do not (Brack et al, 2000, pp34–7). Such leakage would seriously undermine the environmental benefit expected from the climate commitments undertaken by the parties.

To sum up, even if trade measures are primarily introduced for defensive, offsetting purposes – to reduce the costliness of the enforcer's climate efforts – they may also provide incentives for targets and third parties to improve their climate policies. I now turn to how the pattern of regime interplay can affect the likelihood of such improvement.

## Regime interplay and the potency of climate-related trade measures

The fact that international regimes are conceived within specific issue areas, and defined by activity and geographic area, permit their reasonably clear delimitation. This delimitation is in turn important for the analysis of interaction between international regimes – that is, situations in which the content, operation or consequences of one institution are significantly affected by those of another (Stokke, 2000). Sometimes, such regime interplay implies that the content of a regime is changed, as in cases where components of one institution are used as a model for emulation by actors negotiating another. The compliance system of the Montreal Protocol, for instance, frequently serves as a reference for discussion in other environmental regimes.[11] In other instances the regime itself remains unaffected by the other institution, whereas its ability to shape the behaviour of target groups is enhanced or hampered. Thus, liberal trade rules have the general effect of constraining the use of trade measures as a compliance-inducing mechanism under international environmental agreements (Neumayer, 2001).

Three dimensions of interplay between the climate and trade regimes are drawn upon in this analysis: normative compatibility, participatory interplay, and the form and extent of linkage between the regimes. *Normative compatibility* refers to the degree to which the content and ordering of the regimes in question are compatible. High compatibility between two regimes indicates that rules which address the same activity and geographic area are either substantively consistent or hierarchically ordered. Much of the political interest in institutional interplay originates in a belief that many issue regimes have evolved in isolation from each other and have developed rules which are, under closer inspection, substantively inconsistent and without any clear ordering (Young, 2002). Normative hierarchy can be *de jure*, i.e., based on customary or treaty based rules of pre-eminence (Wolfke, 1993), but it can also be based on differences in the broader institutional capacity of the regimes involved. For instance, the binding nature of one regime, or the financial resources or compulsory dispute settlement procedures it makes available, may lead participants in another regime to defer to the first and seek, in their operation of the other regime, to avoid any conflict (Stokke, 2001).

A second dimension, *participatory* interplay between regimes, refers to the sets of actors they involve. The typical situation is that the memberships of two regimes partially overlap,[12] and the exact pattern is significant in part because it may differentiate between states in terms of rights and obligations. Among WTO members, as discussed below, states which are also parties to the Kyoto Protocol may enjoy a different type of protection from climate-related trade measures than will non-parties.

A third dimension of regime interplay, *linkage*, refers to whether participants in either regime deliberately attend to the relationship between the regimes and seek to influence the extent or way in which one affects problem solving under the other (Stokke, 2001). Cross-institutional coordination aimed at reducing any disruptive effects or maximizing synergies is only one of several forms; autonomous adaptation within one or both of the regimes is another (Oberthür and Gehring, 2001). The linkage dimension highlights the dynamic element of the interplay – i.e., the fact that regimes are operated by social actors who might be able to redress the normative compatibility or participatory pattern, should this be conducive to attaining regime goals.

The following sections will examine how these dimensions affect the potency of climate-related trade measures by asking how they influence the likelihood that such measures will be used and the costs incurred by their targets. The section on normative compatibility will focus on the content and ordering of the trade-related provisions in the two regimes. The analysis of participatory interplay examines how such normative compatibility differs depending on the participation of the target of trade measures in the trade and climate regimes. Finally, the linkage discussion will review the efforts that are made under the two regimes to improve the interplay between them, and how various ways to structure this linkage are likely to affect the potency of climate-related trade measures.

## *Normative interplay: Requirements for compatibility*

The WTO, an intergovernmental organization established in 1995 with a membership of 146 states and customs territories, administers and operates a number of multilateral trade agreements with the overarching objective of liberalizing world trade.[13] Environmental concerns are more strongly articulated in this regime than under its predecessor, based on the 1947 General Agreement on Tariffs and Trade (GATT). The Preamble to the WTO Agreement recognizes the need to make 'use of the world's resources in accordance with the objective of sustainable development' and to 'protect and preserve the environment'.

The Committee on Trade and Environment was established to consider salient issues on the trade and environment interface. Other WTO bodies also regularly attend to this interface, especially the committees and councils which administer agreements that can be relevant to environmental regimes – including those on Technical Barriers to Trade (TBT), Sanitary and Phytosanitary Measures (SPS), Subsidies and Countervailing Measures (SCM), the General Agreement on Trade in Services (GATS), and Trade Related Aspects of Intellectual Property Rights (TRIPS). These bodies, as well as the General Council, operate the WTO agreements on a daily basis, whereas a Ministerial Conference of members convenes roughly every second year. A compulsory and binding dispute settlement system, which may authorize bilateral trade sanctions, adds a judicial dimension to the global trade regime.

Certain key principles under GATT, reproduced in subsequent multilateral trade agreements, are designed to discipline the use of international trade sanctions. The principles of national treatment (Article III) and of most favoured nations treatment (Article I) prohibit discrimination in trade among members. 'Like products' are required to be treated identically with respect to internal taxes and regulations, regardless of their domestic or foreign origin, and regardless of the environmental or other policies of their country of origin.[14] In addition, Article XI provides a general ban on quantitative restrictions. The preference in multilateral trade agreements for market-based measures (such as tariffs) over regulatory measures (such as quotas or embargos) is due to the higher transparency of the former and the narrower scope they provide for discriminatory application and asymmetric access to information.[15]

The 'environmental window' of the global trade regime is provided by certain general exceptions to these various principles, laid out in Article XX of GATT and largely emulated in other WTO agreements. Subject to the chapeau requirement that they 'are not applied in a manner which would constitute a means of arbitrary or unjustifiable discrimination... or a disguised restriction on international trade', trade measures may be compatible with the global trade regime if they are 'necessary to protect human, animal or plant life or health' (paragraph b) or 'relating to the conservation of exhaustible natural resources if such measures are made effective in conjunction with restrictions on domestic production and consumption' (paragraph g).[16] Subsequent decisions by dispute settlement panels and appellate bodies have clarified and developed the ramifications of these exceptions, which are also relevant to climate-related trade measures.

The first step when assessing the WTO compatibility of a discriminatory trade measure is to decide whether it falls within the range of policies explicitly mentioned in the Article XX exceptions. Dispute settlement bodies have taken a liberal view on this question, and there is much to suggest that climate measures, because they aim to protect an enforcer's domestic environment, would fall within both the (b) and the (g) sets of policies (Fox, 1996, p2335). There was no disagreement, for instance, in a case over the lawfulness of import restrictions on gasoline that 'the policy to reduce air pollution resulting from the consumption of gasoline was... within the range of those... mentioned in Article XX(b)'.[17] Another dispute settlement body recently noted that the term 'exhaustible natural resources' in paragraph (g) should be interpreted 'in the light of contemporary concerns of the community of nations about the protection and conservation of the environment', adding that 'it is too late in the day to suppose that Article XX(g)... may be read as referring only to the conservation of exhaustible mineral or other non-living resources'.[18]

The second step is to decide whether the trade measure is either 'necessary to' or 'relating to' the policy objectives that are shielded by the Article XX exceptions. This step has presented enforcers with much more difficult hurdles than the first. The 'necessity' test associated with paragraph (b) has been used to inquire whether the enforcer has exhausted the range of non- or less trade-restrictive policy measures that could be expected to achieve the same objective, before introducing a measure contrary to other WTO provisions. A common, and therefore notable, example of non-restrictive measures which should be tried first is good faith attempts to engage the target in multilateral environmental cooperation. The 'relating to' test of paragraph (g) has in practice been an examination of whether the design and application of the trade measure renders plausible a claim that it is 'primarily aimed at' the conservation of exhaustible resources.[19] Enforcers which have stumbled over this hurdle have designed or applied their measures, including those on trade, in ways that affect foreign producers more severely than domestic producers. This has been taken by dispute settlement bodies as evidence that the primary aim has been protectionism.[20]

Paragraphs (b) and (g) have, in addition, implied what could be termed a 'sovereignty' test, which requires that the trade measure, in conjunction with domestic restrictions, can be reasonably expected to deliver the policy objective without excessive intrusion into the jurisdictional autonomy of the target state with respect to its territory and nationals.[21] The sovereignty test is particularly relevant in cases in which the enforcer justifies trade restrictions not on the basis of characteristics of the product itself, but rather on the process and production method (PPM) by which the product has been fabricated, as would be a case of levies based on climate policy differentials among the home countries of producers. According to the latest authoritative statement on the matter, some such intrusion is inevitable and implied also in non-PPM based restrictions, but a measure is excessively coercive if the target, in order to avoid restrictions on trade, must accept the enforcer's policy without adaptation to the conditions prevailing in the target country.[22]

The third and final step when determining whether an environmental trade measure is compatible with WTO rules focuses on the chapeau requirements that the measure should not be an 'arbitrary or unjustifiable discrimination' or a 'disguised restriction on international trade'. As reasoned by the Appellate Body which provided 'the first... authoritative interpretation of the chapeau' (Schoenbaum, 1997, p275), the purpose of this introductory clause is to avoid the abuse or illegitimate use of the exceptions available in Article XX.[23] The practical test has been to inquire whether any discrimination between domestic and foreign industries, or between different exporting countries, is explicitly justified and unavoidable.[24] Internationally mandated environmental trade measures could form one such justification; this is one reason why unilateral trade measures without any basis in climate regime provisions or decisions will probably have a rougher ride in the WTO dispute settlement apparatus than coordinated measures (Chambers, 2001, p94).[25]

Certain features of the climate regime are rather unhelpful for parties that wish to make use of the WTO's environmental window to introduce climate-related trade measures. First, the constitutive documents of the climate regime confirm the basic principles of the multilateral trade regime.[26] The Framework Convention reminds states that '[m]easures to combat climate change, including unilateral ones, should not constitute a means of arbitrary or unjustifiable discrimination or a disguised restriction on international trade' (Article 3.5), a formulation taken verbatim from the chapeau of GATT's Article XX. Similarly, the Kyoto Protocol requires that parties 'shall strive to implement policies and measures under this Article in such a way as to minimize adverse effects... on international trade' (Article 2.3). No counterbalancing phrases are included similar to those found in the Cartagena Protocol on Biosafety.[27] Second, unlike the ozone regime or instruments under the Basel Convention on hazardous waste, the Marrakesh Accords do *not* provide that the Compliance Committee can mandate or authorize parties to impose trade controls on non-parties or non-compliers. During the negotiation of the Kyoto Protocol, the possibility of including trade measures in the enforcement portfolio was indeed considered, but rejected.[28] An important reason is that controls on trade in climate-related commodities with non-parties or non-compliers, or more limited measures like the application of duties or taxes, would be far more consequential for world trade flows and welfare than are those introduced in other issue areas.[29] Compared to the ozone regime, for instance, the number of products that would be relevant for control is much higher, as is their share of international trade and the difficulty of finding ready substitutes (Brack et al, 2000, pp133–4).

Had trade measures been included in the set of consequences applied by the Enforcement Branch of the Compliance Committee, certain rules under international customary law would have been relevant to the discussion of normative hierarchy, notably that more recent and more specialized rules take precedence over earlier and more general ones.[30] Since the Kyoto regime, as specified in the Marrakesh Agreements, is the more recent of the two, and any explicit provision for climate-related trade measures would have been more specific than the general WTO rules, the *lex posterior* and *lex specialis* rules could have been invoked in an argument that climate rules take precedence among Kyoto parties. With no such provisions in place, and climate regime rules that explicitly confirm the general principle of the multilateral trade system, such an argument would be most difficult to defend.

## *Participatory interplay*

Whatever the compatibility of trade measures with various normative commitments, their potency as policy instruments is higher the more asymmetric the interdependence between enforcer and target (Knorr, 1977). If the enforcer depends upon the target economically or politically, the risk of retaliation makes trade measures much less likely. Similarly, if a target of restrictions can easily find alternative and equally well priced markets, trade measures are unlikely to influence its climate behaviour.[31] It follows that the probability of trade measures, and also their force as external enforcement mechanisms, is higher the broader the coalition of sanctioners and the smaller their dependence on the target state for market or other purposes. It has been noted already that the legal implication of the silence on trade measures in the Kyoto enforcement portfolio is to weaken any claim that parties have implicitly accepted such measures among themselves. Because provisions for authorized or mandated trade measures would have enhanced the saliency of this policy option among parties to the Kyoto Protocol, a political implication of this silence is to complicate any effort to build a broad coalition of sanctioners.

The memberships of the Kyoto Protocol and the global trade regime overlap to some extent, which means that the normative interplay between these regimes will differ between various categories of states. This is because the substantive and procedural rules that apply to the use of trade measures differ depending on whether or not enforcer and target states are parties to the respective regimes. This in turn influences the structural relationship between the states involved. As portrayed in Table 7.1, at least four categories of states warrant attention.

**Table 7.1 Participatory regime interplay: Four categories of targets of climate-related trade measures**

| | | Participation in Kyoto Protocol | |
|---|---|---|---|
| | | **Non-party** | **Non-complying party** |
| **Participation in WTO** | **Non-member** | I | II |
| | **Member** | III | IV |

First, a few developing countries which are non-members of both the WTO and the Kyoto climate regime, including carbon-intensive Algeria and Iraq, are shielded neither by the non-discrimination provisions in WTO agreements nor by their weaker echo in the Kyoto Protocol. Any institutional limitation on the right of potential enforcers to target this category of states for climate-related trade measures would stem from other regimes.[32] As long as only industrialized countries have assumed limitation commitments, such a shield is probably unnecessary. However, when the time is ripe for some developing countries to pledge limitation of GHG emissions for the second commitment period or later, the group of non-members of both regimes will be highly vulnerable to being targeted for climate-related trade measures should Kyoto parties find it in their interest to do so.

A second category of states, including Russia should it decide to ratify the Kyoto Protocol, are industrial countries committed under Annex B of the Protocol but which are not members of WTO. Should the Enforcement Branch conclude that parties in this category have exceeded their assigned amounts of emissions in the Kyoto Protocol's first commitment period, the consequences that are *internal* to the regime will definitely be costly. Notably, these parties would be suspended from eligibility to make transfers under Article 17, they would be required to prepare Compliance Action Plans, and their assigned amount for the second commitment period would be reduced by a quantity corresponding to their excess emissions multiplied by a factor of 1.3.[33] On the other hand, the uncertainty and inaccuracy of the underlying information, and the double majority required in the Enforcement Branch for decisions on whether to proceed with a case after a preliminary examination, provide ample opportunity for a party that may be in 'scientific' non-compliance (see Chapter 4 by Berntsen and associates) to escape the internal enforcement mechanism altogether.[34] And if, despite all this, internal enforcement measures are employed, the costs associated with these measures are very much future costs that may prove insufficient to influence present practices. For instance, the deterrent effect intended by the multiplier on excess emissions will be reduced if some parties, anticipating their own non-compliance in the first commitment period, take into account the expected deduction of tonnes penalty from their assigned amounts in the second commitment period when developing their negotiating position over second period targets.[35] Confidence in the potency of internal Kyoto compliance mechanisms will also be diminished if there are doubts over whether the state targeted for response action will remain within the Kyoto regime (see Chapter 5 by Hagem and Westskog).

In these cases where internal means of compliance fail, in part or completely, the introduction of unilateral or coordinated trade measures may present a more attractive policy option, and the target's non-membership of the WTO implies that multilateral trade rules would not interfere. True, as parties to the Kyoto Protocol, this second category of countries would receive some protection from Article 3.5 of the Protocol. This protection is rather limited, however, as there is no established practice within the climate regime on how to interpret 'arbitrary or unjustifiable discrimination', and no compulsory and binding dispute settlement system to help create such a practice.[36] Conversely, WTO rules would not discipline the target of trade measures in its search for an appropriate retaliation, implying that the structural relationship between enforcer and target will be the main factor considered.

A third category of states in this regime interplay are WTO members that are industrialized but not committed to emissions limitations under the Protocol, such as the US. Should such states fail to introduce equivalent or similarly costly climate mitigation measures as the Kyoto committed states,[37] it would be of interest to compare their legal situation with that of countries comprising a fourth category: WTO members that are parties to the Kyoto Protocol but fail to comply with Protocol commitments. Both non-parties and non-compliers are partially shielded by the non-discrimination rules of the global trade regime since, as noted, there is little to indicate that the Kyoto regime harbours any suggestion that climate objectives should take precedence over free trade. Werksman (1999, p2) argues that trade-related disputes among Kyoto parties should be settled by climate regime bodies, but differences between the two regimes in terms of institutional capacity would suggest otherwise. The specific, streamlined and compulsory dispute settlement apparatus of the global trade regime has only a generally framed and non-compulsory counterpart in the climate regime. By implication, a WTO member that perceives its trading rights as violated will likely find the trade regime to be the preferred instrument. Therefore, whether or not the target of trade measures is a Kyoto party, the enforcer must be prepared to defend its case on the basis of the Article XX exceptions.

The key question, accordingly, is how the findings above on the normative interplay between the climate and trade regimes are influenced by whether the potential target belongs in the third or the fourth category in Table 7.1. Because both the particular design and application of a trade measure frequently determine its compatibility with multilateral trade rules, any general discussion of this question is bound to be inadequate. Some elements of an answer can be provided, however. We may assume, for instance, that whatever the status of the target under the Kyoto Protocol, any climate-related trade restriction will survive the first step of a compatibility assessment and be accepted as falling within the range of policy measures intended in paragraphs (b) and (g). As for the 'necessity' test, which is part of the second step of such an assessment, an enforcer targeting a non-party can refer to its decade-long and futile effort to engage that state in a cooperative emissions mitigation programme which aims to include all industrialized countries. Similarly, unlike a Kyoto-committed target with a poor compliance record, a non-party cannot submit that the internal compliance system, which is not restrictive of cross-border flows of goods and services, should have been applied first and allowed to work for some time before introducing trade measures. Failure to try less trade-distorting measures is one of the most frequent complaints by dispute settlement bodies when having found trade restrictions non-compatible with WTO rules. This might also affect reasoning under the 'relating to' test, because the availability of untried alternative measures that could plausibly obtain the policy objective calls into question whether the trade restriction is 'primarily aimed at' environmental protection.[38] The 'necessity' and 'relating to' tests could therefore prove to be easier to pass if the target is a non-party to the Kyoto Protocol rather than a non-complier.

For its part, the sovereignty test would address whether the trade measure can reach its objective – climate mitigation – without unduly interfering with the jurisdictional autonomy of the target state. On the face of it, such interference is greater, and therefore even less acceptable, if the target is a non-party to the Kyoto Protocol, since non-compliers have assumed commitments and thus already reduced their leeway with regard to climate policy. On the other hand, even parties to the Protocol have retained large flexibility as to their choice of policies and measures for meeting their climate commitments, including the flexibility mechanisms. It follows that whether or not the target is a Kyoto party, climate-related trade measures would probably fail the sovereignty test if the target is not offered a similar range of avenues to avoid restrictions on trade. Given that the quota price provides a non-arbitrary standard that Kyoto parties can refer to when deciding upon domestic mitigation efforts, the enforcer would probably have to convince the WTO Dispute Settlement Body that the costs, if any, incurred by the target's policy on climate change are less than those corresponding to the quota price.

The third step, inspection of the chapeau requirement, largely turns on whether the design and application of the trade measure discriminate against foreign producers in ways that are avoidable or unjustified. There is little to suggest that such an inspection will be influenced by the status of the target under the Kyoto Protocol.

To sum up, while the first and third steps of a WTO compatibility assessment would not differentiate between non-parties and non-compliers, the 'necessity' and 'relating to' tests of the second step would be more difficult to pass if the target is a non-complier. This is because unlike non-parties, non-compliers have been at least partly responsive to less trade-restrictive policy instruments and thus are less legitimate targets of trade measures.

## *Linkage interplay*

The linkage aspect of regime interplay highlights deliberate efforts to enhance synergy between regimes or avoid normative tensions. Existing linkages between the trade and climate regimes are rather limited. There is some direct contact in that the Framework Convention Secretariat has observer status in the WTO Committee on Trade and Environment and frequently attends the so-called 'information sessions' held in conjunction with Committee Meetings, where trade-related measures under environmental agreements are presented and discussed at a general level.[39] However, its presentations are usually short and simply provide information about recent developments under the climate regime.[40] Given that no instrument under the Framework Convention includes any reference to trade measures, it is not surprising that the climate regime has been rather peripheral in these discussions.[41]

Accordingly, climate–trade linkages have largely been played out by autonomous adaptation rather than by coordination between the respective regime agencies. In both the climate and the trade regimes, there has been considerable preparedness among participants to accommodate the objectives of the other regime. This is most explicit on the climate side of the relationship, as shown in the incorporation of WTO language in the provisions which address trade,[42] and by not including trade measures among the consequences to be considered by the Enforcement Branch of the Compliance Committee. Within the trade regime, a corresponding policy has emerged as part of a general eagerness to manage the relationship with environmental agreements in a way which can avoid direct shoot-outs between environmental and trade provisions.[43] Early on, the Committee on Trade and Environment discussed a proposal that the WTO Subsidies and Countervailing Measures Agreement be reviewed to ensure that it does not shield governmental subsidies of production that could harm the objectives of the Framework Convention, especially with respect to energy use.[44] In the 2001 Doha Declaration, which created the agenda for the new Millennium Round of trade negotiations, ministers agreed to launch negotiations on how global trade rules are to apply to WTO members that are also parties to environmental agreements with trade obligations.[45] Among the proposals that have been tabled in talks under the Committee on Trade and Environment, several could affect the potency of climate-related trade measures, whether unilateral or coordinated under environmental regimes. Only some or parts of the proposals below will be negotiated in the Millennium Round, however, since the Doha mandate is limited to 'the relationship between existing WTO rules and specific trade *obligations* set out in multilateral environmental agreements',[46] which appears to preclude negotiators from explicitly addressing provisions under environmental agreements that only authorize trade measures.

- Increased use of the *temporary waiver instrument* provided in GATT Article IX for environmental purposes has been suggested and could be used to shield climate-related trade measures. Most commentators find this option politically unrealistic, difficult to operate over time given the size of the WTO membership, and generally vulnerable to challenge (Schoenbaum, 1997, p283; Sampson, 2001, p80).
- A process-oriented adaptation proposal with some coordinative potential is to establish a *consultative mechanism,* which would provide WTO and environmental agencies with a practical and flexible instrument to avoid disputes over trade-related environmental measures.[47] By itself, such a mechanism would hardly affect the potency of climate-related trade measures but it would allow more specific communication between trade and environmental diplomats, which is broadly viewed as necessary to manage the relationship between trade and environmental concerns more effectively.[48]
- A more far-reaching option is to *amend the GATT agreement* itself, for instance by explicitly permitting states to differentiate between products on the basis of PPM characteristics when these are important for the realization of environmental goals (Jenkins, 1996, p226–7. Another amendment could be to insert, as has been done in the North American Free Trade Agreement, a permanent specific

waiver implying that certain multilateral environmental agreements take precedence over WTO rules. The political feasibility of this option is hampered by the requirement of a two-thirds majority and the fact that such an amendment would only apply to those parties that have accepted it (Neumayer, 2001, pp177–9).

- The most realistic option, therefore, is an *interpretive statement* by the WTO Ministerial Conference or General Council on the Article XX exceptions aimed at reducing the room for inconsistency between various panels and appellate bodies (Neumayer, 2001, pp145–6). The impact of such a statement on the potency of climate-related trade measures would of course depend on the substantive interpretation. The general trend in dispute settlement practice is that the various tests, especially the 'necessity' and 'relating to' tests, are becoming easier to pass. On the other hand, this trend is controversial in many quarters, especially the Appellate Body decision in the 'shrimp case', which opens the door for PPM-based trade measures (Bhagwati, 2002; Charnovitz, 2002b). There is always a possibility, therefore, that those critical of this trend will use the Millennium Round to push for a return to a more restrictive application of the environmental exceptions, even in cases that involve coordination under multilateral environmental agreements.

Adaptations within the climate regime can also influence the potency of climate-related trade measures. Some commentators believe that trade measures are bound to play a major role in future amendments to the Kyoto Protocol (Neumayer, 2001, p162). The explicit inclusion of trade measures among the responses to non-compliance under the Kyoto Protocol would enhance the prospective effectiveness of trade measures in several ways, depending on the status of the target. If trade measures against non-compliers are considered, such a provision would reduce the risk that the measure will be challenged under WTO; and if enforced, it would add to the immediate costs incurred by the regime's internal compliance system. Even if the target of trade measures is a non-party to the Kyoto Protocol, such a provision would be significant because it would indicate to a dispute settlement panel or appellate body that no arbitrary or unjustifiable discrimination occurs between parties and non-parties with comparable climate performance.

Such inclusion of trade measures is hardly for the near future, however. The Marrakesh Accords are impressive in specifying the broad compliance provisions in the Kyoto Protocol, but the Compliance Committee has not yet constituted itself or embarked upon its work. While trade negotiators find themselves on a scheduled track to address the agenda agreed upon in Doha, those who operate the global climate regime are unlikely to invest much energy at present in discussing whether the range of consequences to be considered by the Enforcement Branch should be expanded.

## Conclusions

Climate-related trade measures implemented against non-parties to the Kyoto Protocol or parties which fail to meet their commitments will most likely be presented as defensive and offsetting. An important impact, however, may be to render participation in and compliance under the climate regime more attractive. This chapter has examined how the potential of such measures to induce more climate-friendly policies is affected by three aspects of interplay between global trade rules and the Kyoto climate regime.

At the core of this institutional interplay is the *normative* consistency of the trade-related rules of the two regimes and the hierarchical relationship between them. The stronger clout of the WTO compliance system and the compulsory nature of its dispute settlement procedures suggest that should a member of this regime believe that its rights are violated by climate-related trade measures, the normative compatibility of the two regimes would be settled by WTO bodies. Such bodies have so far tended to interpret narrowly the exceptions to the general ban on embargoes and discrimination. The deference of the climate regime to the basic principles of the WTO, and the lack of any provisions for authorization or coordination of such measures under the climate regime, make it probable that this would be the case in a climate-related dispute as well.

This said, the nature and significance of normative compatibility depends upon the *participatory* interplay of the two regimes, especially how they differentiate groups of actors with regard to the rights and obligations they enjoy under the regimes when seen in conjunction. Non-members of the WTO receive the least protection, and their vulnerability will largely be determined by their interdependence relationships, economic and political, with the prospective enforcers. Such power considerations will weigh heavily among WTO members as well, but *ceteris paribus*, trade measures that are authorized or preferably mandated under a multilateral environmental agreement would generally fare better than unilateral ones, especially if the target of trade restrictions is party to the environmental agreement. Beyond this, the findings of a dispute settlement panel or appellate body would presumably differ depending on the status of the target under the Kyoto Protocol. A non-complier with Kyoto commitments would be more shielded than a non-party, because by joining the Kyoto regime a non-complier has exposed itself to regime-internal and less trade-intrusive measures that should be exhausted first.

The final aspect of interplay discussed here, the *linkage* of rule making and decision making under the trade and the climate regime, may in principle change this situation. Today there is only moderate cross-agency coordination, but considerable attention within each regime to the desirability of avoiding conflict between them. Participants in the global trade regimes have decided to renegotiate the balance struck between protection against unjustified trade discrimination and room for pursuance of environmental objectives. The mandate for the Millennium Round of trade negotiations could result in modifications of the WTO regime which would expand the environmental window of the global trade regime, for instance by an interpretive statement on the environmental exceptions which confirm or even broaden the limited deference to environmental goals afforded in recent dispute settlement body reports. On the other hand, the lack of trade provisions in the Kyoto regime, and the fact that further strengthening the contents of the Protocol is not a current preoccupation of climate negotiators, make it less likely that provisions or concerns articulated under the climate regime will be prominent in this WTO process.

## Notes

1 An abridged version of this chapter has appeared in *International Environmental Agreements* vol 4, no 4, 2004, pp341–360. Very helpful comments from Steve Charnovitz, M. J. Mace, Arild Moe, Sebastian Oberthür, Kristian Tangen, Morten Walløe Tvedt, Davor Vidas and my fellow contributors to this book are appreciated.

2 See Marrakesh Accords, Decision 24/CP.7, Sec XV, Arts 4–5; reproduced in FCCC/CP/2001/13/Add.3. On the notion of internal and external compliance mechanisms, see the Introduction to this book and Chapter 6 by Hovi.

3 In a list of environment-related cases brought before WTO compiled by Neumayer (2001), the US had introduced the contended measure in six out of twelve cases; see also WT/CTE/W/53/Rev.1, Annex 1. Many of the unilateral trade measures that are introduced go unchallenged, however.

4 See the more detailed discussion of this below.

5 WT/CTE/W/191, 6 June 2001, Compliance and Dispute Settlement Provisions in the WTO and in Multilateral Environmental Agreements.

6 For a treatment of the broader relationship, including the WTO compatibility of Annex-I country trade in permits, environmental subsidies and fuel taxation, see Brack, 2000.

7 Note that in climate politics, the key enforcer role would have to be adopted by parties other than the US, which has been the most inclined to assume this role in other environmental issue areas , since this country has so far rejected quantified climate commitments.

8 See the exchange on this between Pape, 1997, and Baldwin, 1999.

9 See the strong emphasis of this 'positive' and 'non-punitive' rationale in the Ozone Secretariat's presentation to the Committee on Trade and Environment, WT/CTE/W/57, pp6–7.

10 On border tax adjustment for environmental purposes, see Charnovitz, 1994, especially pp498–505.

11 See Chapter 10 by Wettestad.

12 A domestic aspect of participatory interplay between regimes, not pursued here, is the relationship between the sectors of government they affect, such as trade, environmental or energy production. See Oberthür and Gehring, 2001.

13 Agreement Establishing The World Trade Organization, Art III; all WTO agreements are available online at www.wto.org; membership data as of 4 April 2003.

14 Certain exemptions to the non-discrimination principle are related to regional customs unions and developing countries.

15 WT/CTE/W/47, 2 May 1997, Taxes and Charges for Environmental Purposes – Border Tax Adjustment.

16 A third environmental window is provided by para (d), which exempts measures which are 'necessary to secure compliance with laws or regulations which are not inconsistent with… this Agreement'. In cases involving environmental objectives, this exemption has in practice been subordinated to (b) and (g) in that measures that are found not to be justified by the latter have been deemed 'inconsistent with… this Agreement'; see WT/CTE/W/53/Rev.1, 26.

17 Gasoline, Panel Report, para 6.21, cited in WT/CTE/W/53/Rev.1, 12; see also Werksman, 1999, p10.

18 The report is cited in WT/CTE/W/53/Rev.1, 17.

19 The 'primarily aimed at' phrase was introduced by the dispute settlement panel in the US–Canadian Salmon/Herring Case; see WT/CTE/W/53/Rev 1, pp18–19.

20 See WT/CTE/W/53/Rev 1, pp18–22.

21 See WT/CTE/W/53/Rev 1, pp13–16 and pp22–6. Note that this is not an 'effects test', since substantiation of an actual causal effect is not required. What is required is that the measure be designed to make such a positive effect plausible; see Gasoline, Appellate Body report, pp20–2, cited in WT/CTE/W/53/Rev 1, 23.

22 See Shrimp, Appellate Body report, para 121.

23 Gasoline, Appellate Body Report (1996), p25, cited in WT/CTE/W/53/Rev 1, p7.

24 See WT/CTE/W/53/Rev 1, pp7–11.

25 This is also reflected in Art 31 of the mandate for the Millennium Round of trade negotiations, which refers to the relationship with specific trade obligations under multilateral environmental agreements and not to environmental purposes more broadly.

26 On deferential, collaborative and autonomous linkages between international institutions, see Leebron, 2002.

27 The Biosafety Protocol, adopted in 2000 and available at www.biodiv.org/biosafety, states in its Preamble that 'trade and environment agreements should be mutually supportive' and while '[e]mphasizing that this Protocol shall not be interpreted as implying a change in the rights and obligations of a Party under any existing international agreements', including the WTO, it makes clear that 'the above recital is not intended to subordinate this Protocol to other international agreements'.

28 The Montreal Protocol compliance system, in which trade measures figure prominently, served as point of reference for negotiators of the Kyoto Protocol (Oberthür and Ott, 1999, p216, p282). When reporting on the negotiations of the Protocol to the Committee on Trade and Environment a few months prior to the adoption of the Protocol, the Climate Secretariat noted that speculation on the 'potential trade-related developments or trade issues' of the outcome would be 'premature', WT/CTE/W/61, 16 September 1997, p1.

29 See the discussion by Charnovitz, 2003, pp156–7.

30 The *lex posteriori* rule is codified in Art 30 of the 1969 Vienna Convention on the Law of Treaties (United Nations Treaty Series, 1155); for a discussion of this rule as well as the *lex specialis*, see Wolfke, 1993, pp94–5.

31 Note that the measure could still be perceived as successful by having offset in at least some markets any competitive advantage associated with climate measures, and thus reduced or prevented industrial leakage from less to more emission-intensive production.

32 Unlike Algeria, Iraq is also a non-party to the Framework Convention.

33 Marrakesh Accords, Decision 24/CP.7, Annex Sec XV, Art 5.

34 See Chapter 3 by Mitchell and Chapter 4 by Berntsen and associates (both on the nature and impacts of uncertainty) and Chapter 2 by Ulfstein and Werksman (on Enforcement Branch procedures). Note also the argument by Hagem and Westskog in Chapter 5 that countries such as Russia, which are big sellers of assigned amounts under the Protocol, fossil fuels or energy-intensive products, are less likely than small countries to be targeted for internal enforcement because Enforcement Branch members may take into consideration the economic impacts on their home countries that would result from price effects in those markets.

35 Negotiations on the second commitment period are expected to be finalized before compliance is ascertained with regard to the first period. This expectation may prove wrong, however, or a party may have sufficient information to anticipate that its Kyoto target will be hard to reach.

36 Framework Convention, Art 14.

37 Both the Montreal Protocol and several international fisheries regimes recognize in their provisions for compliance-related trade measures that such measures will not be applied to so-called 'cooperating' states, i.e., non-parties who voluntarily adhere to the substantive commitments.

38 Further evidence would be sought by examining whether the overall set of measures taken in pursuance of climate goals affects foreigners more negatively than national producers. However, there is little reason to expect that this latter examination would differ systematically depending on the status of the target under the Kyoto Protocol.

39 Trade and Environment News Bulletins, TE/029, 30 July 1999.

40 See for instance WT/CTE/W/201 (2001), WT/CTE/W/174 (2000) and WT/CTE/W/61 (1997).

41 Summaries of recent information sessions are given in TE/025 (1998), TE/029 (1999), TE/33 (2000), TE/34 (2000), and TE/36 (2001).

42 See the discussion above; this consistency is highlighted by the Secretariat when presenting to the WTO Committee on Trade and Environment recent developments under the Framework Convention; see WT/CTE/W/61.

43 This concern precedes the establishment of the WTO; on the work of the Group on Environmental Measures and International Trade (the EMIT group) under the Sub-Committee on Trade and Environment, see WT/CTE/W/4 (1995). The group was established already in 1971 but was dormant until around 1990 (Neumayer, 2001, pp10–11).

44 TE004/ 14 August 1995, pp8–9.

45 Doha Declaration, Arts 31–33; available at www.wto.org.

46 Emphasis added in quotation.

47 The Secretariat of the Montreal Protocol has complained about the unwillingness of WTO bodies to provide advice on whether a proposed trade measure would be compatible with global trade rules; see WT/CTE/W/57, 6.

48 In particular, trade diplomats have complained that environmental agencies show insufficient interest in CTE proceedings (see TE/025, 10 August 1998); since then, environmental agency participation appears to have increased.

## References


Baldwin, D. A. (1999) 'The Sanctions Debate And The Logic Of Choice', *International Security*, vol 24, 80–107

Bhagwati, J. (2002) 'Afterword: The Question of Linkage', *American Journal of International Law*, vol 96, 126–34

Brack, D. with Grubb, M. and Windram, C. (2000) *International Trade and Climate Change Policies*, Royal Institute of International Affairs and Earthscan, London

Chambers, B. W. (2001) 'International Trade Law and The Kyoto Protocol: Potential Incompatibilities' in Chambers, W. B. (ed) *Inter-linkages: The Kyoto Protocol and the International Trade and Investment Regimes*, United Nations University Press, Tokyo, 87–118

Charnovitz, S. (1994) 'Free Trade, Fair Trade, Green Trade: Defogging the Debate', *Cornell International Law Journal*, vol 27, 459–525

Charnovitz, S. (2002a) 'The WTO's Problematic "Last Resort" Against Non-Compliance', *Aussenwirtschaft*, vol 57, 409–39

Charnovitz, S. (2002b) 'The Law of Environmental "PPMs" in the WTO: Debunking the Myth of Illegality', *Yale Journal of International Law*, vol 27, 57–110

Charnovitz, S. (2003) 'Trade and Climate: Potential Conflicts and Synergies' in *Beyond Kyoto: Advancing the International Effort Against Climate Change*, Pew Center on Global Climate Change, Arlington, VA

EU Commission, DG Trade (2001) 'Trade and the Environment: Support Sustainable Development', www.europa.eu.int/comm/trade/index_en.htm

Fox, S. T. (1996) 'Responding to Climate Change: The Case for Unilateral Trade Measures to Protect the Global Atmosphere', *Georgetown Law Journal*, vol 84, no 2, 499–542

Friedheim, R. (ed 2001) *Towards a Sustainable Whaling Regime*, University of Washington Press, Seattle

Goldberg, D. L., Van Dyke, B., Bullen, S., Lacosta, N., Muffett, C. and Bartenhagen, E. (1997) 'Effectiveness of Trade and Positive Measures in Multilateral Environmental Agreements: Lessons from the Montreal Protocol Agreements', CIEL, www.ciel.org

Jacob, T. (2001) 'US Industry and Climate Change', DuPont Senior Adviser, Global Affairs, memo, limited circulation by Tom.Jacob@USA.dupont.com

Jenkins, L. (1996), 'Trade Sanctions: Effective Enforcement Tools', in Cameron, J., Werksman, J. and Roderick, P. (eds.) *Improving Compliance with International Environmental Law*, Earthscan, London, 221-228.

Joyner, C. C. and Tyler, Z. (2000) 'Marine Conservation Versus International Free Trade: Reconciling Dolphins with Tuna and Sea Turtles with Shrimp', *Ocean Development and International Law*, vol 31, 127–50

Knorr, K. (1977) 'International Economic Leverage and its Uses' in Knorr, K. and Trager, F. N. (eds) *Economic Issues and National Security*, Regents Press of Kansas, Lawrence, KA, 99–126

Leebron, D. W. (2002) 'Linkages', *American Journal of International Law*, vol 96, 5–27

Neumayer, E. (2001) *Greening Trade and Investment: Environmental Protection Without Protectionism*, Earthscan, London

Oberthür, S. and Gehring, T. (2001) 'Conceptualizing Interaction between International and EU Environmental Institutions', Ecologic, Berlin, www.ecologic.de/interaction/890_deliverables.html

Oberthür, S. and Ott, H. E. (1999) The Kyoto Protocol: International Climate Policy for the 21st Century, Springer, Berlin

Pape, R. A. (1997) 'Why Economic Sanctions Do Not Work', *International Security*, vol 22, 90–136

Sampson, G. (2001) 'WTO Rules and Climate Change: The Need for Policy Coherence' in Chambers, W. B. (ed) *Inter-linkages: The Kyoto Protocol and the International Trade and Investment Regimes*, United Nations University Press, Tokyo, 69–85

Schoenbaum, T. J. (1997) 'International Trade and Protection of the Environment: The Continuing Search for Reconciliation', *American Journal of International Law*, vol 91, 268–313

Stokke, O. S. (2000) 'Managing Straddling Stocks: The Interplay of Global and Regional Regimes', *Journal of Ocean and Coastal Management*, vol 43, 205–34

Stokke, O. S. (2001) *The Interplay of International Regimes: Putting Effectiveness Theory to Work*, FNI Report 10/2001, The Fridtjof Nansen Institute, Lysaker

Werksman, J. (1999) 'Greenhouse Gas Emissions Trading and the WTO', *Review of European Community and International Environmental Law*, vol 8, 1–14

Wolfke, K. (1993) *Custom in Present International Law* (second edition), Martinus Nijhoff, Dordrecht

Young, O. R. (2002) *The Institutional Dimensions of Environmental Change: Fit, Interplay, and Scale*, MIT Press, Cambridge, MA

# Part IV

## Compliance, NGOs and International Governance

Chapter 8

# The Role of Green NGOs in Promoting Climate Compliance[1]

Steinar Andresen and Lars H. Gulbrandsen

## Introduction

There has been a tremendous growth in the number of non-governmental organizations (NGOs) participating in international negotiations and conferences about environmental issues during the last two decades. Most scholars agree that NGOs do make a difference in global environmental politics (e.g. Chatterjee and Finger, 1994; Princen and Finger, 1994; Wapner, 1996; Raustiala, 2001), but it is contestable to what extent their efforts have actually affected international negotiation outcomes and domestic implementation of commitments. Little of the literature on NGOs has addressed what actual influence they have on policy outcomes. Studies that have addressed this question have often confused influence with NGO access, activities or resources in assessing policy outcomes (Betsill and Corell, 2001).[2] Thus, a systematic approach to measuring and analysing NGO influence has been called for (Betsill and Corell, 2001).

In this chapter, we address NGOs' role in promoting climate compliance under the Kyoto Protocol. Because the Kyoto Protocol is not yet in force, we will focus on how the NGOs have contributed to the design of the compliance system and what their options are for enhancing future compliance. Although we do not pretend to measure NGO influence and causality patterns systematically, we make an informed evaluation of their role in promoting climate compliance. More specifically, the following three questions have been singled out:

1 What strategies have NGOs used to influence the development of the compliance system?
2 Have NGOs succeeded in influencing the design of the compliance regime and closely related issues?
3 What possibilities, both internal and external to the compliance regime, are there for NGOs to promote compliance with climate change commitments?

The first two questions can essentially be answered by studying the development up to the present. The last question is more forward-looking and therefore also more speculative and general. Also, we do not define compliance in a narrow legal sense as this would exclude actors such as the US, which will not become a party to the Kyoto Protocol. We conceive of compliance in the broader political sense, and we explore what opportunities there are to promote the climate performance of both parties and non-parties to the Kyoto Protocol. In the following, the three questions above will be addressed in turn.

## Different NGOs: Different strategies, resources and targets

A first observation is that non-state actors in the climate process not only include environmental groups, but also research and academic institutes, business and industry associations, labour organizations, religious bodies and consumer groups.[3] Although our focus is on the green NGOs, this is not a homogenous group. On the one hand we have the traditional activist groups; on the other hand, there are more 'pure', research-based groups with the legal and/or technical expertise to promote environmental goals.

On this basis, we distinguish between activist organizations, which obtain funding and legitimacy through membership and popular support, and advisory organizations, which obtain funding and legitimacy through their ability to give policy recommendations and provide decision makers with legal, technical or scientific advice. NGOs that are clearly activist are Greenpeace and Friends of the Earth (FoE). Important advisory organizations or 'think-tanks' include the Center for International Environmental Law (CIEL), Environmental Defense (ED), the Foundation for International Environmental Law and Development (FIELD) and several others.[4] The World Wide Fund for Nature (WWF) arguably belongs to both categories.

Against this backdrop we differentiate between two main strategies.[5] First, an NGO can pursue an *insider strategy,* seeking to attain influence by working closely with negotiators and governments by providing policy solutions and expert advice. There are many US-based groups in this category. They also engage in knowledge construction, producing research-based reports and papers on particular topics (Gough and Shackley, 2001, p338).

Second, NGOs can pursue an *outsider strategy* promoting compliance with international agreements by putting pressure on negotiators, governments and target groups through campaigning, letters of protest, rallying, direct actions, boycotts and even civil disobedience. The tactic here is to influence public opinion in order to induce states to be more flexible in international negotiations, to push governments to comply with international commitments, and to give polluters and environmentally harmful corporations negative public exposure.

Although the insider–outsider dimension is likely to vary among NGOs, several environmental organizations, especially the major ones with large resources, are likely to pursue a dual strategy. Global activists like Greenpeace and WWF also engage in knowledge construction, using scientists and analysts to acquire further understanding on complex issues. The increasing complexities of many international environmental issues, not least the climate issue, have necessitated this dual strategy. Advisory NGOs, however, usually rely on the insider strategy only.

The broad insider/outsider categories can be broken down in terms of what arenas (actors/institutions) the various types of NGOs target. We assume that NGOs seek to influence one or some combination of the following four arenas, depending upon the type of NGO:

1 *International negotiations and processes*: In our case, this involves efforts to promote a strong compliance system during the negotiating process. All green NGOs generally participate as observers during the various negotiating sessions. This channel is particularly important for the think-tanks, feeding ideas into the negotiating process, while pressure and various mechanisms of 'shaming' are more important for the activist NGOs. Most green NGOs participating in the climate change negotiations are united in the Climate Action Network (CAN). The major NGOs also have considerable independent international activities that take place outside the framework of CAN.
2 *Domestic climate policy and ratification*: This arena is also important for all major NGOs, but in somewhat different ways. The insider NGOs may participate in brainstorming and trying to 'sell' their ideas to their country's delegation and government. Activist NGOs may push for domestic ratification of the Kyoto Protocol and seek to influence the development of domestic climate policy instruments in both parties and non-parties to the Protocol. NGOs such as Greenpeace, WWF and FoE are particularly important in this regard as they have a large number of country offices and can concentrate resources on key countries in the ratification process.
3 *Target groups' climate policy and behaviour*: There are several target groups for climate compliance: oil and natural gas companies, energy industries, transport, industry production involving greenhouse gas (GHG) emissions, agriculture and waste are among the most important. As long as the Kyoto Protocol has not entered into force, and as long as most states have not yet established forceful domestic climate policies, strategies aimed at influencing target groups directly are potentially an important part of the activist NGOs' repertoire. This is likely to continue when/if the Protocol enters into force as behaviour change in target groups is ultimately the only way to reduce GHG emissions.
4 *Public opinion*: This is another important, but diffuse, target for the activist NGOs. They may try to influence public opinion and create awareness to put pressure on governments and target groups. For organizations relying on membership as a significant resource base, this is an important channel – not only to achieve actual influence, but also to attract new members.

This leads us to a final point regarding the potential for NGOs to influence climate policy in general and compliance more specifically: what kind of resources the various types of green NGOs have. There are several sources of leverage, or capital, that NGOs can rely on to transmit information and to influence decision makers, including the following:

1 *Intellectual base*: issue-specific knowledge held by the NGO and its ability to provide decision makers with expert advice and analysis.
2 *Membership base*: the number of members the NGO has, both nationally and internationally.

3 *Political base*: the NGO's access to decision makers and politicians in office.
4 *Financial base*: the financial resources that the NGO can channel into campaigns, lobbying, participation at conferences, commissioning of expert reports, etc.

Several categories could be added to the list, but the purpose is to show that the types of leverage an NGO can apply will contribute to defining the organization's *opportunity set* with regard to gaining political influence. Further, the resources that an NGO has at its disposal are closely linked to the types of strategies it will choose and the arenas it will target. The intellectual base is the prime weapon of advisory NGOs, but other major NGOs are equipped with this tool as well. The more specific expertise and know-how the relevant NGOs have concerning the system of compliance, the higher the potential for influencing the making of the compliance regime. We will lump the other three categories (membership, political and financial base) together and label them 'political clout'. This is relevant mostly to the large activist NGOs. We would assume that the higher the score on this aggregate dimension, the higher the possibility that green NGOs could influence climate policies and thereby increase compliance.

It is important not to confuse resources with actual influence in promoting compliance with the Kyoto Protocol. Resources are characteristics associated with an environmental organization that may or may not translate into political influence. A number of other variables will be decisive as well for promoting compliance. To sum up, two kinds of environmental NGOs will be investigated: activist NGOs and advisory NGOs. These types of NGOs are expected to differ somewhat on three dimensions relevant to promoting climate compliance: resources, levels targeted and strategies. The assumed relationship between NGO type and the three dimensions is set out in Table 8.1.

**Table 8.1 Relationship between NGO type and resources, levels targeted and strategies**

| | **Activist NGOs** | **Advisory NGOs** |
|---|---|---|
| **Critical resource** | Membership base | Intellectual base |
| **Arenas targeted** | International negotiations<br>Domestic policy<br>Target groups<br>Public opinion | International negotiations<br>Domestic policy<br>(Target groups) |
| **Strategy** | Dual strategy: insider and outsider | Insider only |

## Green NGOs and the compliance regime

To measure influence, we will rely on two data sources: *goal attainment*, measured as the correspondence between NGOs' positions and proposals and actual negotiations outcomes; and *ego and alter perceptions*, i.e. how NGOs perceive their influence themselves and how negotiators and other key actors judge their influence. As we have interviewed mostly NGOs, there is a chance that we may overestimate their influence. Also, interviews were conducted with a limited number of actors, but we believe them to be some of the most important ones relating to the compliance issue. The conformity between the environmental community's positions and the actual negotiation outcome as stated in texts and final agreements is one indicator of their influence in the design of the compliance system.

Important determinants of influence might include decision makers' responsiveness and demand for advice, pressure from business and industry groups, and, not least, the potential for alliances with more powerful actors.[6] The problem of causality looms large when trying to isolate the influence of one set of actors from that of others. Within the framework of this chapter we can only deal with this on an ad hoc basis, making an informed evaluation of the different respondents.

### *Access to negotiations and delegations*

While NGOs have been formally accredited as observers to the climate change negotiations since the talks began in 1991, participation in the negotiations have in practice taken the following forms: access to the conference venue, presence during meetings, interventions during debate, the face-to-face lobbying of delegations, and the distribution of documents (Oberthür et al, 2002). Somewhat paradoxically, most of the final negotiations of the compliance procedure, where most delegates agreed on the need for transparency, were conducted behind closed doors (see also Chapter 1 by Werksman). Although participation does not equal influence, it has certainly been a drawback for the green NGOs to have been shut out from important forums. NGOs have therefore had to rely on traditional 'corridor politics' involving face-to-face lobbying and the distribution of documents in the lobby between sessions.[7] There are ways, however, to overcome the problem of lacking access. For example, there has been a rather small but important network of experts on compliance that has interacted frequently. Some of these are official delegates, others are NGOs and academics.[8] Other ways for NGOs to get more direct access to negotiating tables and other closed forums is through participation in government delegations as representatives of civil society constituencies or as expert advisers (Oberthür et al, 2002, p134). For example, FIELD, Greenpeace and WWF have all helped the Alliance of Small Island States (AOSIS) with policy advice and scientific backup in the climate negotiations (Newell, 2000, p143), and FIELD lawyers have frequently been accredited as members of small islands delegations (Oberthür et al, 2002, p135).[9]

Apart from participation and lobbying internationally, access to national governments is crucial for NGO influence. Access to governments can be in the form of consultative meetings and regular meetings with civil servants (Newell, 2000). Many US-based advisory NGOs have worked closely with the government, and they have enjoyed a high degree of access (Eikeland, 1994; Newell, 2000). According to Newell (2000, p132), the World Resources Institute (WRI), Natural Resources Defense Council (NRDC), Environmental Defense (ED), the Woods Hole Research Center and the Audubon Society 'all worked closely with US policy makers and UN agencies in formulating policy options on climate change'. In the US, there were regular meetings before negotiating sessions that started out as meetings between delegations and different kinds of non-state actors. Over time these have been split up, and green NGOs meet separately. US decision makers have said that open brainstorming and other kinds of interaction have been very useful for them (interview with Bodansky, 2000). However, with the change of the US administration, environmental NGOs in the US no longer enjoy a high degree of access to the government, and they have to use other channels to gain influence in future climate change negotiations (interview with Anderson, 2002).

One strategy to increase access, or compensate for lack of access, is to form alliances with other environmental NGOs to share information and coordinate positions. In the climate change issue, nearly all environmental NGOs coordinate their positions through CAN.[10] Created in 1989, CAN is now a global network of almost 300 environmental NGOs working to curb human-induced climate change to ecologically sustainable levels.[11] To achieve this end, CAN members exchange information, work out joint position papers at climate change negotiations and coordinate strategies at the international, national and local levels. As the recognized umbrella NGO in the international negotiations, CAN unites activist and advisory environmental NGOs in one network. CAN is split into a number of working groups according to issue areas, and there is a separate group on compliance issues. Over time, CAN is said to have developed into a well functioning body, characterized by good procedures, open discussions and loyalty by member organizations. (interview with Gulowsen, 2002; interview with Singer, 2002). Although CAN is more important for the less resourceful groups than for the major ones, the CAN network is usually an effective way of communicating NGO positions with one voice during the climate negotiations.

## *Design of the compliance regime*

Compliance has been described as an atypical issue in international climate negotiations, characterized by the small number of actors involved and a lack of strong opinions (interview with Dovland, 2002; interview with Wiser, 2002). Until Bonn (COP-6 part 2, July 2001) and Marrakesh (COP-7, October/November 2001), the compliance issue was low on the agenda during the negotiations. The questions of targets and timetables and the flexibility mechanisms took most of the negotiators' energy. Until the final stages of the negotiations, compliance received scant attention not only from the negotiators but also from most of the NGO community. This, however, provided a window of opportunity for green NGOs with particular expertise and competence (interview with Wiser, 2002).

The first important workshop on the compliance issue was arranged in Vienna in October 1999. At the Vienna meeting, CIEL and WWF presented a joint paper proposing the main elements of a compliance system for the climate regime, based on lessons from other relevant international regimes. CIEL, a small advisory NGO based in Washington DC, associated itself with WWF not only because it agreed on the compliance issue, but also because WWF had much more political clout (interview with Wiser, 2002). These and a few other NGOs with expertise in compliance were allowed to present their views and insights during a series of informal workshops arranged to address the compliance issue. CIEL and WWF jointly introduced the idea of a dual approach to compliance, which refers to a combination of a facilitative body available to assist parties to comply with their commitments and a judicial-type enforcement body (Morgan and Porter, 1999). This basic approach was similar to the US approach to compliance. In a submission before the Vienna workshop, the US called for a separation of the facilitative and enforcement functions.[12] Still, CIEL/WWF and the US had not worked together on this prior to Vienna (interview with Wiser, 2002). The dual approach has been kept and endorsed by all parties in the Marrakesh Accords.[13]

CAN has left most of the responsibility for the work with the compliance system to a small group of experts. Most of the green NGOs have been neither very interested nor knowledgeable on the issue. The Compliance Working Group is the smallest one in CAN, usually composed of some 20 members and mostly chaired by CIEL, illustrating the rather low-key, technical and complex nature of the issue. Among the large traditional activist groups, WWF has been most active on the compliance issue while organizations like Greenpeace and Friends of the Earth have been less involved. To some extent, this reflects a strategic division of labour between the NGOs (interview with Raquet, 2002). Some NGOs work mainly on the mechanisms, some on compliance and some mainly on external issues.

The legal character of any consequences of non-compliance caused much controversy in the negotiations. All the major green NGOs favoured legally binding consequences, but the decision was deferred to the first Conference of the Parties serving as the Meeting of the Parties (COP/MOP) after the Kyoto Protocol's entry into force – seemingly a major setback for the green movement and other 'progressive' forces. Although it was not much emphasized in the media, the US was previously one of the strongest proponents of the legally binding approach.[14] This could be seen in the pressure it put on reluctant colleagues in the Umbrella Group: Russia, Japan, Canada and Australia. The new administration, however, changed policy, and the pressure on the 'gang of four' was gone. Despite the eagerness of the EU and others to conclude the agreement, the reluctant parties succeeded in getting the issue postponed to the COP/MOP.[15] The green NGOs regretted the outcome, but some are pragmatic and downplay its significance, acknowledging that it is difficult to force a country into compliance in real world circumstances.

Obviously, a key point for all green NGOs was to secure maximum openness and public participation in the compliance regime. Up until Marrakesh, it seemed that an open-access compliance regime with strong public participation would emerge. The US had also been the strongest supporter of this approach among the major actors.[16] Once more, the new US administration changed position on what was believed to be an important principle for the US. With this turnabout, the issue lost its importance during the negotiations. The strongest opponent to an open compliance regime, Russia, seized this opportunity. Russia did not want information open to the public and did not want NGOs to submit information.[17] The compliance system eventually became somewhat less transparent and open than the NGOs had advocated, but they were successful in ensuring that NGO observers were allowed to attend Enforcement Branch deliberations and hearings, unless the Branch decides otherwise. Furthermore, NGOs may submit technical or factual information to the Facilitative and the Enforcement Branch, although the Enforcement Branch is only required to rely upon information from 'official' sources. In sum, then, NGO goal attainment on the compliance system was generally quite high.

### *Sinks, flexibility mechanisms and compliance*

Although there has been little contention between the NGOs regarding the compliance regime, narrowly defined, they have not been unified on closely related issues, in particular sinks and the flexibility mechanisms. This is due not least to differences in philosophies regarding the role of the market as a means to reduce GHG emissions in the implementation of the flexibility mechanisms in the Kyoto Protocol. In general, the more markets and the more sinks there are, the easier it will be for most parties to comply with their commitments. The traditional NGO view has been opposed to such a market-based approach, as it would reduce the need for tough domestic actions to reduce emissions. This, however, does not apply to all green NGOs.

ED relies on a market-based approach. According to Newell (2000, p128), ED 'boasts the largest assemblage of scientists, economists and lawyers of any national NGO working on climate change'. In contrast to most expert organizations, ED has quite a large member base. It was also among the main architects behind the US system of tradable sulphur dioxide ($SO_2$) permits, designed and put into operation more than a decade ago (interview with Petsonk, 2002). ED has worked relentlessly to get the negotiators to adopt a similar approach for the international climate regime. Considering its expertise, close interaction with the US administration and political clout, there is reason to believe that it has had an effect on the design of the Kyoto mechanisms, which are mainly a US brainchild (Grubb et al, 1999). In general, ED has sided with the US against the EU in its interpretation of the Kyoto mechanisms.[18]

ED is also among the few environmental NGOs supporting the previous US administration on the interpretation of sinks – the possibility to claim emission credits for carbon stored in forests and soils.[19] At The Hague (COP-6) and in Bonn (COP-6 part 2), most green NGOs argued that sinks should not be included as Clean Development Mechanism (CDM) projects, and that 'additional activity sinks' (Article 3.4 of the Kyoto Protocol) – which can include land management, agricultural practices and forest management – would put the integrity of the Kyoto Protocol at stake.[20] This placed a considerable strain on CAN (as well as negotiators), and ED excluded itself from CAN at COP-6, but is now once more attending the CAN meetings.[21] The agreement in Bonn and the final agreement in Marrakesh (COP-7) include sinks in the CDM, and the liberal interpretation of the 'additional activity sinks' prevailed during the negotiations, against the mainstream NGO position. We can thereby safely conclude that mainstream NGOs have had little influence on this turn of events. However, we cannot conclude that the outcome is a result of ED's influence because the same outcome was promoted by other, very powerful, actors, notably the US.

This position of ED and some other US-based think-tanks, in opposition to that of Greenpeace and other more traditional green NGOs, mirrored the differences in philosophy between the US and the EU in their regulatory approaches during most of the negotiation process. The fact that ED sided with the US on key points has made it somewhat 'suspect' in many green quarters, and Greenpeace and ED have been at loggerheads. Most of the other major NGOs have been closer to the Greenpeace position on issues like sinks and the flexibility mechanisms. ED also argued for no cap on the flexibility mechanisms to ensure cost efficiency, whereas the major NGOs strongly favoured such a cap. The major NGOs were not successful in persuading the delegates to adopt their positions here either.

ED advocated that compliance be built into the rules for emission trading, claiming that this would make a separate compliance system redundant (interview with Petsonk, 2002). Although ED did not succeed on this, compliance was actually built into the rules for emission trading because a party that is not in compliance is not eligible to sell allowances. The Enforcement Branch will have the authority to suspend and reinstate that eligibility.

Two NGOs, CIEL and CAN, had long fronted the idea of a *compliance fund*, provided that it was designed in such a way that real emissions reductions would be achieved (see Wiser, 2001). The idea, however, was captured and given a new meaning after the US, Canada and France at COP-6 in The Hague tried to use it as a way to introduce a price cap on costs into the Protocol – 'a mechanism that could allow countries to comply with their Kyoto targets by paying a discounted fee instead of accomplishing actual emissions reductions' (Wiser, 2001). This turned out to be a critical issue at the session in The Hague. During the negotiations running up to The Hague, CAN had been able to convince several states to endorse the compliance fund idea; now it suddenly found that it had to oppose the idea due to the new meaning given to the issue. The problem was that most EU ministers 'were at best vaguely familiar with it [the compliance fund], recognizing it only as something the green groups wanted' (Wiser, 2001). At a press briefing, CAN declared war on the new compliance fund, and CIEL and other key expert NGOs mobilized their people on compliance to convince EU ministers to reject the idea. In the end, CAN's war on the price cap and voluntary fund and the expert NGOs' efforts to fight the idea bore results. The EU and those Umbrella Group countries that had never been enthusiastic about the fund – including New Zealand, Australia and Japan – decided against the proposal and the idea never went anywhere.

## *Influence of green NGOs*

Where does this leave us in terms of NGO influence? We are in no position to answer this conclusively, but some observations are warranted. In the initial phase it may be that some of the advisory NGOs qualified as *intellectual leaders* (Young, 1991) as a result of their ability to frame the compliance issue in a novel and constructive way. Their specific impact, however, is uncertain, as the US came up with essentially the same approach when the issue surfaced in 1999. Be that as it may, considering both the lack of knowledge surrounding this complex but important issue and the lack of priority given to it by most delegations, there is no doubt that the persistence and expertise of a few advisory NGOs have been important for the making of the compliance regime. In line with Werksman (this volume), 'Their [NGO experts'] consistent support for a strong enforcement mechanism, and their ability to articulate how such a system could work in practice helped to maintain the focus of the negotiations on the need for an effective Enforcement Branch'. Although this does not apply to all green NGOs, but primarily to a few advisory organizations, CAN and the major groups supported the advisory NGOs' work.

NGOs have been quite successful in attaining their goals for the design of the compliance regime. Goal attainment was high in terms of acceptance for the dual approach to compliance, which includes both a Facilitative and an Enforcement Branch, a strong enforcement mechanism, and potentially significant scope for NGO participation in Enforcement Branch deliberations and hearings. However, in their opinion, the compliance regime could have been better in terms of ensuring legally-binding consequences, and the regime became more closed than they had advocated.

There is some discrepancy between *ego* and *alter* perceptions about NGO influence. According to the Norwegian delegation leader, NGOs had relatively modest influence on the design of the compliance regime, whereas US legal experts played a key role (interview with Dovland, 2002). Nonetheless, the close interaction between some NGOs and delegates 'makes it difficult to pinpoint who influences whom' (interview with Dovland, 2002). Apart from bringing their expertise to the negotiating table, maybe their most important channel of influence has been through *alliances* with key actors. The most important ally regarding the compliance regime on questions like the dual approach to compliance proceedings, public participation and legally-binding consequences was the Clinton administration. When NGOs lost access to such allies with the introduction of the second Bush administration, their influence was dramatically reduced. This may indicate that in absence of such allies, their independent influence is modest. Moreover, their influence was fairly high when compliance was coined in more technical-neutral terms in the early phase. When the issue was more polarized towards the end of the negotiations, their influence was reduced.

In the 'big' polarized issues such as the flexibility mechanisms and sinks, the large activist NGOs have been more involved, but their influence has been very modest. The major NGOs lost the major battles over sinks and over the interpretations of the mechanisms. Among the NGOs we have focused on, ED has seemingly been the most successful on these accounts, but the alliance with the US has, of course, been crucial.

## Opportunities for NGOs to strengthen climate compliance

### *Internal strategies: Using the instruments in the compliance system*

It has been maintained that 'where civil society... has specific expertise, its monitoring capabilities can enhance transparency, increase certainty, and promote compliance' (Wiser, 1999, p4). Let us consider some options for NGOs to strengthen climate compliance by using or enhancing instruments in the climate regime: first is participation in compliance proceedings. There will potentially be significant opportunities for NGOs to participate in such processes. Intergovernmental organizations and NGOs will be entitled to submit technical and factual information to the Compliance Committee's relevant branch – that is, either the Facilitative or the Enforcement Branch (Decision 24/CP.7 of the Marrakesh Accords, Section VIII (4)). The Enforcement Branch is only required to rely upon information from 'official' sources, but it will be difficult to ignore reliable information submitted by competent IGOs or NGOs (see Ulfstein and Werksman, this volume). The two branches of the Compliance Committee may seek expert advice (VIII (5)), possibly giving NGOs scope for influence. Furthermore, the information considered by the relevant branch shall be made available to the public, unless the branch, of its own accord or at the request of the party concerned, decides otherwise (VIII (6)).

Second, NGOs may monitor sinks projects and CDM project activities and attend Executive Board CDM meetings.[22] This is essential to prevent misuse of the Kyoto Protocol in general and the flexibility mechanisms in particular. For example, NGOs work to prevent the CDM – which assists developing states in reducing GHG emissions, while assisting developed states to meet their commitments – from being used to finance 'clean' coal power plants. 'Green' or 'clean' coal plants, a contradiction in terms according to the NGOs, refers to plants with lower carbon dioxide ($CO_2$) emissions than existing plants, or merely lower emissions than the average $CO_2$ emissions of existing plants. NGOs will also try to monitor the quality of CDM sinks projects. Moreover, one of the NGO community's greatest fears is that the CDM could be used to finance nuclear energy, or that the Kyoto Protocol in general might be portrayed as an argument in support of building nuclear power plants. How can the NGOs work to avoid such a development? Executive Board CDM meetings are broadcasted on the UN Framework Convention on Climate Change (UNFCCC) website, which may give good overviews and levels of transparency (interview with Anderson, 2002). Similarly, 'CDM watch' and 'Sink watch' are two Internet sites under development, private initiatives to monitor and keep track of the quality of new projects. Such initiatives might become important, but they are still in a very early phase.

Networks like CAN may be effective in exposing 'bad projects', and big NGOs such as WWF, Greenpeace and FoE may develop their own instruments to monitor CDM projects through their international and national networks. NGOs themselves believe that monitoring big CDM projects will become an important instrument to ensure the quality of such projects (interview with Gulowsen, 2002). Even if NGOs are not able to influence the CDM Executive Board, project investors may very well be sensitive to NGO shaming. Merely the threat of NGO shaming may actually prevent investors from engaging in 'bad projects'. However, even though it is possible to monitor some projects, it will probably be difficult to keep track of all CDM projects. NGOs will, to some extent, have to rely on the CDM rules and focus on them (interview with Anderson, 2002).

Third, another loophole in the Kyoto Protocol, as most NGOs see it, is 'hot air' emission trading. This is the potential of Russia and other Central and Eastern European (CEE) countries to sell some of their surplus GHG emission allowances as part of an international trading regime. Due to industrial and economic changes in Russia and the CEE countries since 1990, these countries have received GHG emission budgets far in excess of what they need. Hot air emission trading may thus lead to emissions being significantly higher than they would have been in the absence of such trading. Quotas must be traceable back to the country of origin, and NGOs are likely to try to prevent parties or investors from trading hot air quotas. NGOs already work to convince parties to refrain from using the 'hot air loophole', and they will probably shame parties that buy hot air quotas from Russia and CEE countries. However, as Newell (2000, p151) points out, private interfirm trading removes an element of public oversight, thus reducing the scope for NGO influence.

NGOs fear that, combined, these loopholes could mean there will be no actual reduction in the global GHG emissions, putting the integrity of the Kyoto Protocol at stake. To persuade parties to refrain from using the loopholes, Greenpeace has developed sophisticated computer models with 'loophole analysis', showing country-specific data on the potential consequences of exploiting the loopholes (interview with Raquet, 2002).[23]

Fourth, although Annex I parties are the main targets of NGO attention, NGOs may also be able to influence the performance of non-Annex I parties. The transfer of technology from Annex I parties to non-Annex I parties under the Climate Convention will be administered through the Global Environment Facility (GEF). The World Bank's policy with respect to GEF is in general to include NGOs in the development and implementation of the facility (Princen and Finger, 1994, p19). The Bank meets regularly with large Washington-based environmental NGOs (Princen and Finger, 1994, p6). One specific point of entrance for NGOs in the climate case is the Ad Hoc Working Group on Global Warming and Energy, under the Scientific and Technical Advisory Panel of the GEF (Newell, 2000, p150). NGOs can try to ensure the quality of technology transferred from Annex I to non-Annex I parties, as well as its appropriateness to local circumstances, but it is likely that governments themselves will secure firm control with capital-intensive projects (Newell, 2000, p150).

Finally, the questions of verification and monitoring (Articles 5, 7 and 8 in the Protocol) are extremely complex and boring for the media and the public – a general problem with the issue area (interview with Gulowsen, 2002). The complex and technical nature of verification and monitoring may be a considerable problem with regard to transparency and openness. Most NGOs realise that, although it will be a tremendous challenge for them to make the whole issue area more interesting to the public at large, it is a necessary step to improve future climate performance.

### *External strategies: Options available outside the compliance regime*

The outsider strategy – that is, directing efforts at government, industry and the public – has been used less than the insider strategy in the climate change issue. Greenpeace and other traditional activists have been relatively more active using the insider strategy. The explanation might be that the character of the climate problem and the framing of the issue in society are more important for the choice of strategy than is the characteristic role of the organization (Pleune, 1997).[24] In comparison to several other environmental problems, the climate problem is far more difficult and complex: its causes can be found in nearly all of society; there are no obvious alternatives to the combustion of fossil fuels; there is no one single target group that can be aimed at; and there are no solutions that will alleviate the problem in the short term (Gough and Shackley, 2001, pp330–1).

These observations are by and large confirmed by those activist organizations we have interviewed. National offices, whether those of Greenpeace or WWF, feel that it is very difficult to generate interest among people due to the complexity and diffuse nature of the climate change problem. It has been maintained that Greenpeace in the US has almost given up on promoting the Kyoto Protocol due to a lack of interest and results (Skjærseth and Skodvin, 2001). More specific and understandable problems are much simpler to work with and 'sell'. It has also been claimed that national actions have not been sufficiently linked to the international negotiation process (interview with Gulowsen, 2002). Enthusiasm on the part of NGOs to campaign for the Kyoto Protocol may also have waned because the Protocol is seen as a very weak instrument to reduce emissions. Still, they support the Kyoto Protocol because it's the only game in town, and they believe that the climate negotiations are a learning experience and a process that will continue to move forward (interview with Raquet, 2002).

After the meeting in The Hague, in November 2000, there had been tremendous disappointment in the green NGO community, and many had declared that they would abandon their efforts because of the lack of financial resources and results. Then, on March 13 2001, US President George W. Bush Jr officially rejected the Kyoto Protocol. This created nothing less than an uproar among a number of European nations and the green NGO community. Paradoxically, the Bush statement injected new energy into the process. It is thus claimed that 'the Bush "no" to Kyoto is the only good thing Mr Bush has ever done for climate protection' (interview with Raquet, 2002). As a direct result, the 'Ratify Kyoto Now' campaign was launched by Greenpeace. It appears that the momentum created was rather short-lived (interview with Gulowsen, 2002), and not all observers share the opinion that it had much of an effect (interview with Dovland, 2002). WWF, in February 2002, launched a similar ratification campaign for the Kyoto Protocol, originally running until the World Summit on Sustainable Development (WSSD) in August–September 2002.[25] The causal chain is long and complex with a number of intervening factors, but it may be that NGO activism on this point had some effect on the agreement reached in Marrakesh in November 2001, as well as on Russia's and Canada's statements at the WSSD in Johannesburg that they would ratify the Kyoto Protocol, thereby bringing the Protocol into force.[26]

Being shamed as a laggard can be harmful for a state (even if not a party), corporation or industry. For a state, its reputation in the international community may be at stake. For small countries like Norway this is certainly a potential problem, but large states, particularly the US, do not seem to care much about 'soft' shaming. For business and industry, their reputations among customers and partners may be harmed. This has little or no effect on some companies, but others are more sensitive. (Skjærseth and Skodvin, 2001).

## Conclusions

In evaluating NGO goal attainment, it has proved necessary to differentiate between the narrowly defined compliance regime and linked issues such as the flexibility mechanisms and sinks, as well as between activist and advisory NGOs. Although the compliance regime, from the NGO perspective, could be better in terms of public participation as well as on legally-binding consequences, overall goal attainment was quite high. With regard to *ego* and *alter* perceptions, NGOs perceive, not surprisingly, that they have had more influence than more neutral respondents estimate. The insider strategy seems to have had some success in shaping the compliance regime, especially in the early phase before the issue was more politicized, but the close interaction between some NGOs and delegates makes it very difficult to reveal actual patterns of influence. Apart from providing knowledge and expertise, green NGOs have relied heavily on forming strategic alliances with key actors. When such alliances have broken down, the NGOs were no longer very influential. This may cast doubt over the extent of their independent influence on the design of the compliance regime.

More indirectly related to compliance issues, NGOs differ on their views on the flexibility mechanisms and sinks. Here, some US advisory NGOs have had high goal attainment, whereas the big, activist NGOs have had low goal attainment. This is not to say that the advisory NGOs have necessarily had more influence than the activists. There can obviously be a measurement problem in assessing the effectiveness of the insider strategy versus the outsider strategy. Advisory NGOs typically have policy solutions that are more acceptable to governments or closer to governments' own positions. Activist NGOs, on the other hand, typically have more radical and far-reaching solutions. It is important to note, however, that the activist groups have expended most effort on these more high level political issues, while some expert groups have had a more narrow focus on the compliance issue. In that sense, the latter group has gotten more mileage from their effort than the major NGOs, whose political clout has so far been of limited significance. On this complex issue it may seem that intellectual capital is more important.

As to the question of how the NGOs can enhance future compliance, there are a number of opportunities to adopt internal strategies by making use of the existing regime. High on the NGO agenda is the effort to close the loopholes related to sinks and the flexibility mechanisms. The extent to which action and shaming will make a difference is likely to depend, *inter alia*, on the size and tradition of the country in question, how the facilitative and Enforcement Branch will operate in practice, and the extent to which environment and climate change are important issues among the public. There are also a number of ways in which NGOs may adopt external strategies and work to promote climate performance independent of the compliance system, including promoting renewable energy and energy saving, cooperating with industry and business, shaming climate 'villains', and working on consumer choices. While such strategies have proved useful in other issue areas, expectations should be more modest here. The complexity of the climate issue has made the traditional activist role of the NGOs very difficult. Seemingly, the climate issue is difficult to 'sell' and for people to relate to. NGOs have engaged in a wide variety of actions with and against states, target groups and parts of the public to rally support for more proactive climate polices and for more speedy ratification processes. There have probably been some direct, and more indirect, effects of this work, but there can hardly be said to have been any major breakthroughs. This is a slow, cumbersome and 'heavy' political and economic process. So far, most evidence points more toward adaptation as a main direction for policy efforts rather than strongly curbed emissions. If this is correct, it may be that the insider strategy with expertise and cooperation with target groups will be more effective than confrontation – but politically this may be difficult for some of the NGOs to swallow.

## Notes

1 Another version of this study has been published in *Global Environmental Politics*, vol 4, no 4 (2004), 54–75.

2 Exceptions to this observation include Corell and Betsill (2001), Newell (2000), Arts (1998) and Skodvin and Andresen (2003). While Corell and Betsill focus exclusively on the international level, Skodvin and Andresen argue that the domestic level should also be included.

3 For an overview of the role of the petroleum industry in this context, see Skjærseth, this volume, Chapter 9.

4 Interviews have been conducted with WWF, Greenpeace, CIEL, FIELD and ED. Therefore, most emphasis will be placed on these organizations.

5 See the discussion in Chapter 9 by Skjærseth of strategies pursued by transnsational petroleum companies.

6 It is the rather rare alliance between powerful states and powerful NGOs that can explain the seemingly very strong influence of green NGOs in the international whaling regime. How strong this influence would have been in the absence of this alliance is more uncertain (Andresen, 2001).

7 It appears the lobby is not the only place for lobbying. It has been claimed that the toilets are used as well. So the concept of 'restrooms' may not apply to key delegates...

8 Many of the respondents pointed to this informal network.

9 According to Arts (1998), it is well known that FIELD made the proposed AOSIS Protocol in 1995.

10 Smaller partnerships between environmental NGOs are also commonplace, be they joint actions, policy proposals or initiatives.

11 www.climatenetwork.org

12 The US submission is dated July 30 1999 (FCCC/SB/1999/MISC.12).

13 The CIEL/WWF working paper also suggested a 'screening committee' (Morgan and Porter, 1999), but this idea was not accepted.

14 The US position follows logically from its reputation of keeping agreements due to the requirement to pass implementation legislation.

15 It appears that the new US negotiation team wanted to uphold the position on this issue, as it was an important principle in US policy, but they were given direct orders from the White House to change position.

16 The EU also supported this approach, but with less vigour.

17 Some G-77 countries supported Russia, and it has been claimed that the EU gave in too easily to Russian pressure.

18 The EU later made a U-turn on this issue and has become a firm supporter of the flexibility mechanisms (Christiansen and Wettestad, 2003).

19 The Nature Conservancy as well as NRDC also supported the use of sinks.

20 The 'additional activity sinks' have also been called the 'do nothing' sinks (Begg, 2002, p334).

21 There do not seem to be any clear-cut rules as to who can be a member of CAN. However, the general rule is that national members must support the organization applying for membership, and the applicant must be accepted by all CAN members (interview with Bradley, 2002).

22 In practice, it seems that NGO attendance at Executive Board CDM meetings is limited to observering a live webcast on the UNFCCC website.

23 This kind of computer analysis was actively used by Greenpeace during the climate negotiations to show delegates the consequences of different proposals on the table (interview with Raquet, 2002).

24 Pleune (1997) found evidence confirming this proposition in a study of Dutch environmental NGOs' strategies in relation to climate change.

25 WWF has drawn up a hit list of 25 states that must ratify to turn the Protocol into international law (World Wide Fund for Nature, 2002).

26 The campaign was given high priority by NGOs in the preparations for the WSSD.

## References

Andresen, S. (2001) 'The International Whaling Regime: "Good" Institutions, but "Bad" Politics?' in Friedheim, R. (ed) *Towards a Sustainable Whaling Regime*, University of Washington Press, Seattle, WA

Arts, B. (1998) *The Political Influence of Global NGOs: Case Studies on the Climate and Biodiversity Conventions*, International Books, Utrecht

Begg, K. G. (2002) 'Implementing the Kyoto Protocol on Climate Change: Environmental Integrity, Sinks and Mechanisms', *Global Environmental Change*, vol 12, 331–6

Betsill, M. M. and Corell, E. (2001) 'NGO Influence in International Environmental Negotiations: A Framework for Analysis', *Global Environmental Politics*, vol 1, no 4, 65–85

Carpenter, C. (2001) 'Business, Green Groups and the Media: The Role of Non-governmental Organizations in the Climate Change Debate', *International Affairs*, vol 77, no 2, 313–28

Chatterjee, P. and Finger, M. (1994) *The Earth Brokers: Power, Politics and World Development*, Routledge, London

Christiansen, A. C. and Wettestad, J. (2003) 'The EU as a Frontrunner on GHG Emissions Trading: How Did it Happen and Will the EU Succeed?', *Climate Policy*, vol 3, no 1

Corell, E. and Betsill, M. M. (2001) 'A Comparative Look at NGO Influence in International Environmental Negotiations: Desertification and Climate Change', *Global Environmental Politics*, vol 1, no 4, 86–107

Eikeland, P. O. (1994) 'US Environmental NGOs: New Strategies for New Environmental Problems?', *The Journal of Social, Political and Economic Studies*, vol 19, no 3, 259–85

Environmental Defense (2000) 'Global Corporations and Environmental Defense Partner to Reduce Greenhouse Gas Emissions', news release, 17 October, www.environmentaldefense.org

Gough, C. and Shackley, S. (2001) 'The Respectable Politics of Climate Change: The Epistemic Communities and NGOs', *International Affairs*, vol 77, no 2, 329–45

Greenpeace (2001) 'Corporate America and the Kyoto Climate Treaty', Greenpeace Briefing, www.greenpeace.org

Grubb, M., Vrolijk, C. and Brack, D. (1999) *The Kyoto Protocol: A Guide and Assessment*, Earthscan, London

Kolk, A. and Levy, D. L. (2001) 'Winds of Change: Corporate Strategy, Climate Change and Oil Multinationals', *European Management Journal*, vol 19, no 5, 501–9

Levy, D. L. and Egan, D. (1998) 'Capital Contests: National and Transnational Channels of Corporate Influence on the Climate Change Negotiations', *Politics and Society*, vol 26, no 3, 337–61

Morgan, J. L. and Porter, S. J. (1999) 'Compliance Institutions for the Kyoto Protocol: A Joint CIEL/WWF Proposal', discussion draft, prepared for the Fifth Conference of the Parties, 25 October

Newell, P. (2000) *Climate for Change: Non-state Actors and the Global Politics of the Greenhouse*, Cambridge University Press, Cambridge

Oberthür, S., Buck, M., Müller, S., Pfahl, S., Tarasofsky, R. G., Werksman, J. and Palmer, A. (2002) *Participation of Non-governmental Organizations in International Environmental Governance: Legal Basis and Practical Experience*, Ecologic, Berlin

Pleune, R. (1996) 'Strategies of Environmental Organizations in the Netherlands Regarding the Ozone Depletion Problem', *Environmental Values*, vol 5, 235–55

Pleune, R. (1997) 'The Importance of Contexts in Strategies of Environmental Organizations with Regard to Climate Change', *Environmental Management*, vol 21, no 5, 733–45

Princen, T. and Finger, M. (1994) *Environmental NGOs in World Politics: Linking the Local and the Global*, Routledge, London

Princen, T., Finger, M. and Manno, J. (1995) 'Nongovernmental Organizations in World Environmental Politics', *International Environmental Affairs*, vol 7, no 1, 42–58

Raustiala, K. (1997) 'States, NGOs and International Environmental Institutions', *International Studies Quarterly*, vol 41, 719–40

Raustiala, K. (2001) 'Nonstate Actors in the Global Climate Regime' in Luterbacher, U. and Sprinz, D. F. (eds) *International Relations and Global Climate Change*, MIT Press, Cambridge, MA

Skjærseth, J. B. and Skodvin, T. (2001) 'Climate Change and the Oil Industry: Common Problems, Different Strategies', *Global Environmental Politics*, vol 1, no 4, 43–63

Skodvin, T. and Andresen, S. (2003) 'Non-state Influence in the International Whaling Commission 1970–1990', *Global Environmental Politics*, vol 3, no 4, 61–86

Wapner, P. (1996) *Environmental Activism and World Civic Politics*, State University of New York Press, Albany, NY

Wiser, G. (1999) *Compliance Systems Under Multilateral Agreements: A Survey for the Benefit of Kyoto Protocol Policy Makers (CC99-2)*, The Center for International Environmental Law, Washington, DC

Wiser, G. (2001) *Kyoto, Costs, and Compliance: A Public Interest Lawyer's View of COP-6*, The Center for International Environmental Law, Washington, DC

World Wide Fund for Nature (2002) 'WWF launches ratification campaign for Kyoto climate treaty', press release, 7 February, www.panda.org

Young, O. (1991) 'Political Leadership and Regime Formation: On the Development of Institutions in International Society', *International Organization*, vol 45, no 3, 281–308

## Interviews

Anderson, Jason, Climate Action Network (CAN) Europe, Brussels, 4 July 2002.

Bodansky, Dan, US State Department, Washington, DC, December 2000.

Bradley, Rob, Climate Action Network (CAN) Europe, Brussels, 4 July 2002.

Dovland, Harald, Norwegian Ministry of the Environment, Oslo, 25 May 2002.

Gulowsen, Truls, Greenpeace, Norway, Oslo, 15 May 2002, and personal communication, October 2002.

Petsonk, Annie, Environmental Defense (ED), Washington, DC, 21 March 2002.

Raquet, Michel, Greenpeace European Unit, Brussels, 4 July 2002.

Singer, Stephan, World Wide Fund for Nature (WWF), European Policy Office, Brussels, 4 July 2002.

Werksman, Jacob, Foundation for International Environmental Law and Development (FIELD), Oslo, personal communication, 1 June 2002.

Wiser, Glenn, Center for International Environmental Law (CIEL), Washington, DC, 19 March 2002.

Chapter 9

# Major Oil Companies in Climate Policy: Strategies and Compliance

Jon Birger Skjærseth

## Introduction[1]

ExxonMobil urged the Bush administration in a memo to the White House to replace Robert Watson, Chairman of the Intergovernmental Panel on Climate Change (IPCC) since 1996. Shortly after, President George W. Bush Jr decided to take the advice of the world's biggest private company, and Watson is now no longer IPCC chairman.[2] Besides actively disputing IPCC science, ExxonMobil has exerted political pressure far beyond its headquarters in Irving, Texas: UK Prime Minister Tony Blair was recently forced to defend a meeting with ExxonMobil's Chief Executive Officer (CEO) and Chairman, Lee Raymond, and assure the general public that the UK would not change its climate policy as a result of the meeting.[3] These two incidents reported by international media illustrate the visible tip of the iceberg of corporate power in climate policy. More than 50 per cent of global greenhouse gas (GHG) emissions originate from the activities of multinational corporations, and oil is responsible for about one quarter of the 'greenhouse effect' (Gleckman, 1995). In a broader perspective, a wide range of global environmental problems have been linked to the worldwide activities of multinational corporations, including ozone depletion, loss of biodiversity and species, overfishing and illegal trade in hazardous wastes (Retallack, 2000).

Major European and North American oil companies have chosen significantly divergent paths in their climate policy strategies.[4] The European majors, Shell and BP, support the Kyoto Protocol, while ExxonMobil opposes it. Shell, BP and ExxonMobil are the biggest privately owned oil companies in the world and are now often referred to as the 'super majors', or 'the three sisters'. These super majors are clearly influential since in terms of revenue they are all among the top ten largest companies in the world, irrespective of activity.[5]

This chapter looks more closely at the differences in corporate climate strategies with a specific focus on compliance and compliance systems. On the one hand, any discussion of the role of corporations concerning compliance with the Kyoto Protocol appears far-fetched at this stage, since the Protocol is not yet in force. Moreover, international climate obligations apply to states only, meaning that industry will only be indirectly affected by the domestic implementation of climate commitments. On the other hand, corporations can affect what states agree to and whether states comply with what they have agreed to through domestic and international channels. Companies constitute critical target groups for compliance since they partly control the behaviour that has to be modified: states depend on the active or reluctant cooperation of the non-state actors causing the problem in the first place in order to fulfil their international commitments.

More specifically, major oil companies and other target groups can at least influence at the following:

- international instruments and compliance systems linked to the Kyoto Protocol;
- national and regional instruments and compliance systems developed to implement the Kyoto obligations; and
- state compliance with climate obligations.

US oil companies that oppose the Kyoto Protocol have mainly focused on the Protocol obligations rather than its compliance system. Since the flexibility mechanisms and the compliance system are ways for industrialized (Annex I) countries to meet the emissions targets set by the Protocol, the strategy of the US oil companies has been to prevent US participation and put the entire Protocol out of action. US participation in the Kyoto Protocol in turn has consequences for state compliance with climate obligations, i.e. the ability of the US to meet the aim of limiting emissions of GHGs included in the United Nations Framework Convention on Climate Change (UNFCCC).[6] Participation has been conceived of as a crucial element of compliance since broad participation is important for meeting the commitments of any global agreement (Dannenmaier and Cohen, 2000). European oil companies that support the Kyoto Protocol have been most concerned with national and regional instruments and compliance systems. These efforts are more developed and more imminent than the international system.

Against this backdrop, this chapter focuses on two specific questions. First, to what extent and how have major oil companies affected the US exit from Kyoto Protocol and the design of national/regional instruments and compliance systems linked to the Kyoto obligations? Second, why have the oil majors chosen such different climate strategies related to the Kyoto Protocol? Global oil companies want to sell as much oil and gas as possible at the highest possible price in the same global market; the business opportunities and challenges offered by the problem of climate change would thus apparently be the same for ExxonMobil, BP and Shell. Thus we would expect that their different strategies would have to be explained by other differences, such as their internal structures or political contexts. Identifying the sources of these different strategies may provide knowledge as to whether and how corporate resistance to a viable climate policy can be overcome.

This chapter is structured as follows. The following section discusses corporate influence on the US exit from the Kyoto Protocol and the design of compliance systems and emissions trading in the UK and the EU, which has established one of the first carbon dioxide ($CO_2$) emissions trading systems. The US and the EU have to a large extent determined the international climate commitments since the adoption of the UNFCCC in 1992. The chapter then explores explanations for the observed differences in corporate climate strategies. The chapter ends with some concluding remarks on the implications of these explanatory factors for the future strategies of US companies in a situation without the Kyoto Protocol.

## Corporate climate strategies and influence

The oil industry earns its livelihood from oil, natural gas and coal – the main sources of emissions of GHGs – and will be severely affected by regulatory measures to curb GHG emissions. This section looks at how the oil industry has adopted significantly different strategies on either side of the Atlantic. The US oil industry and the wider US fossil fuel lobby have opposed any binding climate targets and actually influenced the US decision to opt out of the Kyoto Protocol. In Europe, the oil industry, represented by Shell and BP, has not only passively supported the Kyoto Protocol, but has actually led the way towards a viable climate policy by adopting voluntary GHG emissions reporting, verification and emissions trading.

The literature on corporate management and the natural environment tends to distinguish between some notion of *proactive/offensive* and *reactive/defensive* environmental strategies (Roome, 1992; Steger, 1993; Ghobadian et al, 1998).[7] Given a certain environmental risk inherent in a company's activities, a proactive company motivated by profits and survival will exploit new market opportunities and support environmental regulation. Conversely, a reactive company will deliberately leave market opportunities unexploited and oppose regulation. However, corporate actors do not only develop environmental strategies as a response to market opportunities or environmental risk; they also actively seek to influence other corporations, societies, governments and international regimes. The oil majors can facilitate compliance with climate commitments in different ways. First, they can support the Kyoto Protocol, including an effective international compliance system. Second, companies can adopt targets on GHG emissions and implement corporate programmes in order to reach those targets. In contrast to states, a multinational company can require its branch offices around the world to comply with its climate policy. Third, corporations can report in-house GHG emissions and conduct third party verification of emissions data. Fourth, corporations can lead the way towards compliance by setting an example for other corporations as well as governments. Conversely, corporations can obstruct compliance by actively weakening international and domestic obligations and by not adopting any voluntary corporate measures.

The notion of industry influence is, however, difficult to pinpoint precisely because the concept is closely related to power, and power and influence have in turn mainly been related to states (Betsill and Corell, 2001). Power has, in state-centric terms, been related to capabilities, while influence is seen as a relationship between actors that can modify behaviour (Cox and Jacobson, 1973). In the same way, we may say that large corporations possess structural and instrumental capabilities. First, structural influence is related to states' dependence upon industry: industry is important for economic growth, employment and technological innovation, particularly in the energy sector, which tends to be viewed as a strategic state objective. This structural dependency frequently provides industry with privileged and informal access to decision making. Second, instrumental influence is based on huge in-house financial, human and technological resources, which can be deployed to obstruct or support climate policy.

The companies under scrutiny here all possess significant structural and instrumental capabilities, making them capable of influencing climate policy. The important question is thus how corporate capabilities are translated into influence.[8] Here, it might be useful to distinguish between types of corporate climate strategies on the basis of the mechanisms whereby influence is exercised. Two types seem particularly relevant in this context.[9] The first type is corporate influence by means of coercion. This will typically be related to a reactive corporate strategy and express itself through the use of 'sticks and carrots' aimed at manipulating the cost–benefit calculus of decision makers. According to this type, we thus expect companies to lobby decision makers by the use of threats and persuasion aimed at affecting the perception of costs and benefits. Corporations can threaten lawsuits, hire PR firms to put pressure on politicians and develop economic models showing the high public (and consequently political) costs of regulating GHG emissions.

The second type of corporate influence is exercised by means of unilateral action. This tends to be related to a proactive strategy and expresses itself by demonstrating that a certain 'cure' does indeed work, or by setting a good example for others to follow (Underdal, 1991, p142). For example, mandatory corporate GHG emissions reporting and independent verification have been opposed by industry for competitive reasons. A proactive company can lead the way for others by displaying relevant data which are fundamental for compliance with any emissions trading system.

Determining the influence of corporate actors empirically is extremely difficult. First, there are different theories of business–state relations. One view takes the position that the state actively serves business interests that are able to act cohesively in the political arena. Another position argues that the state can maintain neutrality and independence from business interests (Levy and Egan, 1998). While these opposing views can be regarded as empirical propositions that may lead to different answers in different cases, we should bear in mind that the structural influence of industry is very difficult to observe. Moreover, it is difficult to separate the influence of specific corporations or branches from those of other corporations and branches. Thus, we have to take a broader view on industry in general, particularly the fossil fuel industry in which the major oil companies have played an important, if not dominating role.

Admittedly, corporate influence depends upon a number of conditions, including access to decision making, the strength of counterbalancing forces such as the green movement, and the cohesiveness of the strategies. A comprehensive comparative analysis of the relative weight of corporate influence goes far beyond the purpose of this chapter. Instead, the task here will be to search for observable direct influence compatible with the types of mechanisms whereby influence is exercised.

### *Corporate influence on US climate policy*

The US oil industry and the wider fossil fuel lobby have actively opposed binding targets and timetables including any mandatory compliance systems since the adoption of the UNFCCC in 1992. The outcome of the UNFCCC was determined by the US and was not, strictly interpreted, very far from industry interests.[10] Obligations to report were directed towards states, and the Intergovernmental Panel on Climate Change (IPCC) guidelines developed subsequently focused on national activity data rather than data from specific facilities. The convention did not include any specific binding targets and timetables, and it did not restrict the parties' choice of policy instruments. On the other hand, the UNFCCC clearly aimed at limiting emissions of GHGs. Five years later, in December 1997, the Kyoto Protocol was adopted and signed by the Clinton–Gore administration. The Protocol departed significantly from the interests of the fossil fuel lobby and the oil industry by requiring specific and binding reduction objectives linked up to flexibility mechanisms and compliance systems. By rejecting the Kyoto Protocol as a 'fatally flawed agreement' in March 2001, President George W. Bush Jr brought US climate policy back to square one and back in line with the interests of the US fossil fuel industry. The US withdrawal from the Kyoto Protocol threatened and continues to threaten the Protocol and thus has ramifications far beyond US compliance with the provisions of the climate regime.

The power and influence of the US oil industry and the wider fossil fuel lobby is one important reason behind the US withdrawal from the Kyoto Protocol. It is well documented that the US pluralist political system provides ample room for strong interest groups to lobby for their interests in energy and climate policy (Hatch, 1993; Newell, 2000; Kolk, 2001). From 1993 to 2001, the Clinton–Gore administration tried to develop a viable climate policy at home as well as internationally by keeping 'Big Oil' at arm's length while consulting with the green movement. This strategy proved unsuccessful owing to congressional resistance as a result of intensive coercive lobbying by ExxonMobil and the fossil fuel industry. In 2001, George W. Bush Jr took office and the US oil industry changed its strategy from 'sticks' to 'carrots'. Since 1999, ExxonMobil has been one of the most generous political donors supporting Republican candidates in the US. In return, energy officials representing the Bush administration have met with Exxon and other energy industry leaders, while at the same time deliberately excluding the green movement.[11]

From 1989 to 2002, the Global Climate Coalition (GCC) effectively exploited the wide room for lobbying in the US Congress. The GCC has been described as perhaps the most powerful corporate lobby organization in climate policy (Raustiala, 2001). ExxonMobil had a seat on the board and played a leading role within the GCC as well as within the American Petroleum Institute (API) (Skodvin and Skjærseth, 2001). ExxonMobil opposes the Kyoto Protocol; it has not set any reduction targets for its own GHG emissions and does not have any immediate plans to invest in renewable energy. ExxonMobil's CEO and Chairman, Lee Raymond, has become famous for his outspoken criticism about the scientific basis for climate policy.[12]

The GCC represented interests from almost every important sector in US industry, including trade associations and the oil, coal, utility, chemical and auto industries. According to the former head of the US climate delegation, Robert Reinstein, it was obvious that the GCC would get attention when it represented a very significant proportion of the US GDP (Newell, 2000, p100). The GCC spent a serious amount of money to convince policy makers that proposals to limit $CO_2$

emissions were 'premature and ... not justified by the state of scientific knowledge or the economic risks they create' (Levy, 1997, p58). The GCC directed and coordinated massive opposition to the US ratification of the Kyoto Protocol from its tiny office in Washington, DC, a stone's throw away from Congress. For example, the GCC sponsored a study concluding that the Kyoto Protocol would cost the US over 2.4 million jobs and reduce the GDP by US$300 billion annually (Carpenter, 2001).

The GCC was weakened in 1996 when one of the three oil super majors, BP, withdrew from the organization. Shell followed BP in 1998, and a number of US companies such as the Ford Motor Company and Texaco followed the European companies from 1999 onwards. It should be noted, however, that even though the US companies could no longer accept the aggressive anti-climate stance of the GCC, they remained firmly opposed to the Kyoto Protocol even after their exit (Skodvin and Skjærseth, 2001). The GCC closed after 13 years. According to the GCC, the group was deactivated because it had served its purpose by contributing to a new national approach to global warming, i.e. the US exit from the Kyoto Protocol. However, another, and perhaps more important, cause was probably the declining support within the business community following BP's withdrawal, since the Protocol is still alive and future US participation cannot be excluded.

As noted, the GCC and the US oil industry have not been very active on compliance systems related to the Protocol.[13] Since the Kyoto mechanisms and compliance systems are means of meeting the Protocol's targets for Annex I country emissions, the support of oil companies and other corporate actors for these systems varies with their support for the Protocol. Representatives of the GCC were present in the international negotiations on compliance, but they never took the floor.[14] Their strategy has been to cripple US climate policy and put the entire Kyoto Protocol out of action by lobbying against any binding targets and timetables within the US.

The first significant victory of the US fossil fuel lobby dates back to 1993. Shortly after the Clinton–Gore administration took office in 1993, President Clinton presented his economic plan, which included a British thermal unit (BTU) tax based on the heat content of the fuel. When the tax proposal came to the Senate Finance Committee in 1993, it was clear that it could be killed before a full vote in Senate if one Democrat voted against it. The API and a wide range of other fossil fuel industry interests formed the American Energy Alliance to defeat the tax. The alliance hired a public relations firm, Burson-Marsteller, which was able to use the media to put local pressure on Democrats. Senator Boren from Oklahoma was the first to give in, thus sinking the tax proposal (Agrawala and Andresen, 1999). According to the API, much of its resistance to the climate policy of the Clinton–Gore administration can be traced back to the BTU tax proposal, which was perceived as extremely provocative by the fossil fuel industry.[15]

At the first UNFCCC Conference of the Parties (COP) in 1995, the US agreed to the Berlin Mandate, which declared that non-binding commitments for developed countries were inadequate and that no new commitments would be imposed on developing countries. The latter had been approved by Vice President Gore, but was met with sharp criticism from industry lobbyists and the Republican Congress, subsequently forcing the administration to retract its position (Agrawala and Andresen, 1999).

In the period leading up to Kyoto, lobbying activities intensified in the US (Kolk, 2001). For example, the GCC sponsored a US$13 million television campaign that claimed that the price of petrol would increase by 50 cents per gallon if Kyoto timetables were implemented.[16] In July 1997 – five months before the Kyoto meeting – the Senate voted by an impressive 95–0 majority against any treaty that would exempt developing countries from legally binding commitments and imply higher energy costs (the Byrd–Hagel resolution). This non-partisan congressional resistance has been directly linked to the powerful lobbyists representing the GCC in general and the oil and coal industry in particular (Newell, 2000; Agrawala and Andresen, 2001). Nevertheless, the last-minute agreement in Kyoto between the US and the EU apparently violated the terms set by the Congress by exempting developing countries and committing the US to cut its GHG emissions by 7 per cent from 1990 levels by 2008–12. The breakthrough came after Vice President Gore's visit to Kyoto, where he called upon the US negotiators to increase negotiating flexibility.

Since the Senate would have to ratify the Kyoto Protocol, the Byrd–Hagel resolution led the fossil fuel lobby to believe that the Protocol would never be ratified in the US. ExxonMobil, the API and the GCC viewed the Protocol as 'dead on arrival', regardless of the upcoming presidential election in 2000. Democratic candidate Gore would push for ratification, but he had a low standing in the Senate and would face considerable opposition no matter what. George W. Bush Jr would have a higher standing in the Senate, but would not push for ratification. And even if the Protocol was ratified against all odds, the GCC threatened to issue lawsuits against necessary legislation that would make US implementation impossible within the commitment period.[17] With the benefit of hindsight, the US withdrawal two years later should hardly have been surprising.

The reactive climate strategy of ExxonMobil and the US fossil fuel lobby has affected US participation in the Kyoto Protocol and thus the ability of the US to meet the aim of the UNFCCC. The mechanisms whereby corporate influence has been exercised match well with the notion of coercion. Based on huge structural and instrumental capabilities, the US fossil fuel industry has combined sticks and carrots, depending on the position of the ruling US administration: the climate initiatives presented by the Clinton–Gore administration were met with threats of lawsuits, PR campaigns aimed at mobilizing political pressure and economic models showing the high costs of regulating GHG emissions, while the Republicans have received generous political donations.

## *Corporate influence on EU climate policy*

In Europe, the Kyoto Protocol was initially given a cautious welcome by the oil giants. European industry had been most concerned with the EU carbon/energy tax proposed in 1992 and its potential consequences for European competitiveness. The shift in focus to flexibility mechanisms took care of the most controversial aspect of EU climate policy. European industry generally favours the Kyoto Protocol and mandatory compliance systems. UNICE (Union of Industrial and Employer's Confederation of Europe) – representing 34 industry and employer's federations from 27 European countries – supports the Kyoto Protocol and emissions trading and states that 'a fundamental requirement for trading is an effective international compliance regime. An effective compliance regime will be simple, predictable, transparent and equitable.' (UNICE, 2000)

BP and Shell accept the problem at hand as defined by the IPCC. In 1997, Shell's Group Managing Director Phil Watts maintained that the oil industry has 'the privilege of being part of the solution' to this problem (Watts, 1997). The same year, Shell declared that the Group's coal assets were to be sold out (completed in 2000), and Shell established a fifth core business – Shell International Renewables – with an investment plan of US$0.5 billion over the next five-year period. Both Shell and BP explicitly support the Kyoto Protocol and have adopted specific targets for in-house GHG reductions: Shell, for example, aims to reduce its GHG emissions by 10 per cent from their 1990 levels by 2002. Both companies have – in addition to investing in energy efficiency, decarbonization of fuels and renewable energy – launched internal GHG emissions trading systems in order to reach their goals. BP launched a pilot system in 1998 that was extended to cover all 150 business units worldwide in 2000. Shell initiated its STEP (Shell Tradeable Emissions Permit) System in 2000, which is currently used by businesses representing 30 per cent of the GHG emissions from the Shell Group's operations.

As discussed below, these pioneering initiatives have had consequences for a number of compliance-related issues. BP and Shell have voluntarily led the way by setting an example in corporate GHG reporting and verification as well as by developing emissions trading schemes that have served as a model for the UK and the EU.

*GHG reporting and verification*

To ensure that the numbers that governments use to demonstrate compliance are real, GHG emissions reporting and independent verification are critical. Moreover, mandatory reporting and verification is a precondition for emissions trading. The IPCC has developed guidelines for national GHG emissions inventories. These inventories use national activity data rather than data from specific facilities. The system is based on common standards using common source, sink and fuel categories in six major sectors. There are, however, important links between national and corporate systems as to the types of GHGs included, the methodology used for estimating emissions and baseline years (Loreti et al, 2001).

BP and Shell have been instrumental in developing the Greenhouse Gas Protocol Initiative, aimed at promoting internationally accepted GHG accounting and reporting standards for companies. Both companies have contributed to the development and testing of the standards. In 1998, the World Business Council for Sustainable Development (WBCSD) and the World Resources Institute (WRI) developed the protocol initiative. The protocol is based on a multi-stakeholder partnership of business, NGOs and governments (World Resources Institute, 2002).

BP and Shell have also been the frontrunners in this field at the US Pew Center (see the concluding section) by verifying GHG emissions data by third parties (Loreti et al, 2001). The purpose is for each company to demonstrate progress toward its own GHG emissions targets, support emissions trading and build stakeholder confidence. Reports on GHG emissions as well as the verification results are available at their websites. In the case of BP, verification was conducted by several consulting, verification and financial auditing firms. These efforts were in turn supported by an independent expert panel including green groups, representatives from governments, academia and the UN (World Resources Institute, 2002).

A growing number of companies have shown interest in emissions verification as a consequence of BP's and Shell's activities. Even ExxonMobil has decided to report carbon emissions.[18] The example set by BP and Shell is important since mandatory GHG reporting and disclosure is seen as the first essential step to addressing climate change. Moreover, these companies have shown the way for public authorities. Mandatory reporting on GHG emissions has led to a heated debate between Democrats and Republicans in the US Senate related to the proposed energy bill.[19] In the EU, comprehensive mandatory reporting of annual environmental and social reports – extending far beyond climate policy – has been proposed for companies with over 250 employees. This initiative has, however, met resistance from EU business organizations arguing that European firms are too diverse to have a single reporting requirement.[20]

*Emissions trading in the UK and the EU*

Reporting, verification and the consequences of non-compliance constitute crucial elements of one of the first $CO_2$ emissions trading systems in operation. The scheme forms part of the UK Climate Change Programme, which was launched in November 2000 and aims to go well beyond the UK's commitment to reduce GHG emissions by 12.5 per cent below 1990 levels under the EU's burden-sharing arrangement for Kyoto compliance. Shell and BP have been instrumental in developing the scheme by leading the way in pioneering in-house trading systems, and both companies have taken on significant reduction targets within the scheme (Table 9.1). It is quite illustrative that Chris Fay, who previously headed the Royal Dutch Shell subsidiary Shell UK, has been appointed by the UK authorities to promote the trading scheme. No US-based oil companies with operations in the UK participate.

The trading scheme is voluntary and open to all corporations operating in the UK. At an auction held in March 2002, 34 companies took on emissions reduction targets in exchange for a financial incentive worth UK£215 million over the five years of the scheme. The auction was considered a success: emission reduction targets for 2006 added up to more than four million tonnes of $CO_2e$ (carbon dioxide equivalent), representing about 5 per cent of the planned savings between 2000 and 2010 under the UK Climate Change Programme. On average, the targets represent an 11 per cent reduction from companies' baseline emissions.[21] Participants can cut GHG emissions in-house, or by buying or selling carbon allowances.

**Table 9.1 Oil industry participants in the UK emissions trading scheme[22]**

| | Reduction target for 2006, tonnes $CO_2e$ | Incentive payment, £1000s | Gases other than $CO_2$ |
|---|---|---|---|
| BP | 353,500 | 18,866 | Methane |
| Shell UK | 438,750 | 23,416 | Methane |

All 34 participants are required to report emissions in order to demonstrate compliance with their targets. Detailed reporting guidelines on principles, annual emissions and baselines must be followed by the corporations. These guidelines draw upon the IPCC procedures and practice. Moreover, emission inventories must undergo a process of verification before being reported to the UK government. The UK Accreditation Service (UKAS) has drawn up a list of verifiers and will be assisting the verifiers before granting accreditation.

Compliance is defined as sufficient allowances to cover total emissions during the previous compliance period for each participant. Participants must thus be able to demonstrate at the end of each annual reconciliation period that they are in compliance. During the initial stage, non-compliance (i.e., the failure to hold sufficient allowances) will have the following consequences: the non-payment of the financial incentive, and a reduction in the number of allowances allocated for the next compliance period. The UK government intends to introduce legislation after Parliament has accepted the compliance regime as legally binding. When the statutory compliance regime is in place, non-compliance will – in addition to non-payment and the reduction of allowances – lead to a fixed financial penalty for each tonne of $CO_2e$ emitted for which a participant did not hold an allowance (DEFRA, 2001). Thus, voluntary participation in the UK emissions trading scheme indicates that corporations supporting emissions trading also accept mandatory reporting, independent verification and a legally binding compliance regime that includes penalty sanctions. BP and Shell have explicitly warned against any voluntary trading system, arguing that voluntary systems will lead to low stability and predictability.[23]

In 2000, the EU established the European Climate Change Programme (ECCP) to 'drive forward EU efforts to meet the targets set by the Kyoto Protocol' (ECCP, 2001, p3). With its multi-stakeholder approach, the ECCP serves both as an instrument for ensuring progress in the implementation of EU climate policies, and as a vehicle for participation by industry in the process. The EU has adopted an emissions trading directive as the core element of the ECCP. BP and Shell have acted as 'key drivers' in the development of the EU emissions trading scheme (Christiansen and Wettestad, 2002, p12).

The emissions trading directive pays due attention to common monitoring, reporting, verification and enforcement. European industry generally supports the directive. According to the Secretary General of the EU employers' confederation, UNICE, 'This provides European business with the certainty it needs to begin planning for emissions trading which starts in January 2005'.[24] 'Non-compliance is to be dealt with by a financial penalty that is sufficiently high to ensure that it makes no sense for an operator not to go out and buy from the market a sufficient number of allowances to cover the installation's actual emissions' (COM, 2001, p14).

The European Petroleum Industry Organization (EUROPIA) is composed of 21 oil companies and represents the downstream activities (including refineries) of the oil industry within EU institutions. In the early 1990s, EUROPIA was one of the most aggressive opponents of the climate policy of the EU, particularly the carbon/energy tax (Skjærseth, 1994). Today, EUROPIA welcomes the EU emissions trading directive as a learning exercise towards an international system under the Kyoto Protocol. EUROPIA also supports a 'sound and transparent' monitoring system, but makes penalties for non-compliance conditional upon whether the system is binding or voluntary.[25] There are two main reasons underlying the change in the position of EUROPIA: first, the change in the climate policy of the EU and its Member States; second, the change in the strategies of Shell and BP.[26] UNICE supports the proposed EU directive as well as a 'learning by doing' process in the period before 2008. In a detailed position paper on the EU proposal, UNICE had no comments on monitoring, reporting and verification requirements. Concerning enforcement, UNICE supported financial penalties, but recommended changing the wording in the proposed directive to avoid unnecessary uncertainty (UNICE, 2002).[27]

The proactive strategy of BP and Shell has affected emissions trading and compliance systems in the EU. The mechanisms whereby influence has been exercised can be understood in terms of the companies' ability to exert influence, mainly by taking unilateral action that served to set an example for other corporations as well as public authorities. BP and Shell have exploited new market opportunities in renewables and have voluntarily taken on a leading role in corporate reporting, verification and emissions trading.

## Explaining different corporate climate strategies

The previous section showed that North American and European oil companies have chosen significantly different climate strategies. ExxonMobil has chosen a reactive strategy, while BP and Shell have chosen a proactive strategy. The consequences of these strategies extend far beyond rhetoric. The US oil and fossil fuel industries have influenced US climate policy since the early 1990s and contributed to the US exit from the Kyoto Protocol. Conversely, the European oil industry has voluntarily led the way towards, in relative terms, a viable policy aimed at cutting GHG emissions. Even though this 'beauties and beast' distinction between US-based and European-based companies does not necessarily say much about their actual behaviour in terms of GHG emissions, it represents a clear indication of the kind of climate policy futures the oil companies are preparing for.[28]

As noted in the introduction, the oil industry is a global industry operating in a global market. The business opportunities and challenges offered by the problem of climate change would thus apparently be the same for ExxonMobil, Shell and BP. Why then have these companies chosen such different climate strategies? There are at least two different answers located at three levels of analysis. First, different strategies may be a result of differences between the companies themselves. Second, differences may have resulted from the different political contexts in which the companies operate. Differences in regulatory pressures and market opportunities may be generated at the national level as well as the international. International regimes and the EU can generate different responses among corporate actors to the extent they generate different rules, norms and principles for different (groups of ) actors.[29]

These two explanatory perspectives have some important consequences for governance in the field of climate change. If different climate strategies result mainly from differences between the companies themselves, the room for affecting corporate strategies by political means tends to be limited. Conversely, if political context accounts for the observed differences, the outlook for political institutions to influence the strategies of multinationals is brighter.

With respect to corporate-specific factors, the business management literature has developed a broad spectrum of factors to explain corporate environmental strategy choice.[30] Factors such as the environmental risk faced by the companies, learning capacity, environmental reputation, leadership, corporate tradition and ownership have been singled out at the corporate level. Even though previous analyses have identified some differences along these lines, they have concluded that corporate differences do not sufficiently explain the significant differences in climate strategies (Rowlands, 2000; Skjærseth and Skodvin, 2001). Similarities rather than differences dominate the picture of company-specific features related to economic and organizational factors relevant to climate strategies. ExxonMobil, Shell and BP are highly comparable in terms of core business areas, exploration and production volumes, resource reserves and incomes. Moreover, the three companies currently have very similar organizational structures – that is, they are all based on a corporate headquarters governing all parts of the organization according to the same principles, which ensures communication across all divisions. The most important difference between Shell, BP and ExxonMobil is perhaps the relative importance of European and US markets, which provides a backdrop for exploring the impact of factors related to the political context in these regions.

The 'nationality' of private multinational companies appears to be particularly important for their attitudes and strategies (Rowlands, 2000; Skjærseth and Skodvin, 2003). Thus, in this context, the companies' home countries – the locations of their historical roots, their headquarters and many of their activities – are the main sources of explanation for strategy choice. The basic assumption is that corporations are affected by the social demand for environmental protection, the governmental supply of relevant policy, and the political institutions that link this supply and demand.

The variation in social demand for climate policy and the governmental supply of such policy create different pressures and opportunities for companies. A high degree of social demand creates the risk of consumer campaigns and boycotts of petroleum products like gasoline, which can affect the companies' market shares. On the other hand, a high degree of social demand also creates opportunities: companies can respond to green consumers' willingness to pay higher prices for clean energy products. Likewise, an ambitious climate policy in terms of targets and policy instruments exposes companies to regulatory pressure, while an active public policy on renewables creates market opportunities. A high degree of social demand for climate policy combined with an ambitious climate policy thus appears to be a forceful (and logical) combination in democratic systems stimulating proactive strategies. Conversely, a low degree of social demand and a lenient climate policy may provide corporations with correspondingly scant incentives (Skjærseth and Skodvin, 2003).

The political institutions that link corporations and states can also influence the strategies of corporate actors. However, corporations do not only represent political targets for governmental policy, they also represent in themselves a social interest group with the potential to influence governmental policies. On the other hand, governmental decision makers can, in stereotypic terms, organize this relationship as a conflict-oriented process where the state imposes regulations on corporations excluded from the decision-making process, or as a collaborative, inclusive process where the state consults and negotiates goals and policy instruments with target groups. The conflict-oriented approach rests on punishment in cases of non-compliance and aims to avoid regulatory capture (where the regulated takes control over the regulator). This strategy is likely to produce resistance and a reactive strategy among companies. Conversely, the aim of a consensus-oriented approach is to raise awareness and promote social responsibility among companies. This strategy appears more compatible with a proactive corporate strategy. Whereas these approaches are likely to spur different corporate strategies, their relative short-term impacts on environmental policy effectiveness are not obvious.

It thus appears that a high degree of social demand, the supply of an ambitious policy and consensual political institutions will lead to a proactive climate strategy. This is borne out by the cases of the US and the EU (the UK and the Netherlands), which differ systematically along these dimensions, as will be discussed in more detail in the following section.

## *The political context of oil companies*

Shell, BP and ExxonMobil operate all over the world, but they are still firmly tied to Europe and the US, respectively. Shell International is located in the UK, but the company has its backbone in the Netherlands (the Shell Group is 60 per cent Dutch-owned) and has been influenced by Dutch culture, society and policy. BP has its backbone in the UK, while ExxonMobil has deep roots in the US, which can be traced back to Rockefeller's foundation of Standard Oil in 1882. Exxon's cultural heritage has been seen as generally important for the company's choice of environmental strategy (Estrada et al, 1997). This also implies that Shell and BP are much more closely linked to the climate policy of the EU than ExxonMobil.

In relative terms, the US and the EU have experienced different levels of social demand. First, a number of studies have shown that Europeans have been much more receptive to proactive measures on climate change than the US (see, e.g., Skolnikoff, 1997; Rowlands, 2000). The Netherlands is widely perceived to be among the greenest countries in the world in terms of public values and attitudes towards environmental problems.[31] Shell, for example, has exploited this situation by using the Netherlands as a testing ground for the society's willingness to pay for environmental protection in general, and clean energy in particular.[32] High social demand for clean energy will make Shell's investments in renewable energy more profitable. In contrast, the US public expresses significant concern for the environment, but climate change is given little public attention compared to domestic environmental problems (OECD, 1996; Saad and Dunlap, 2000). Public opinion polls have also indicated that the US population is not willing to accept significant costs or a large share of the international burden to reduce the problem (Gallup and Saad, 1997). While Shell perceives consumers' willingness to pay for clean energy as a business opportunity, fossil fuel interests in the US have launched PR campaigns to highlight the high economic costs of reducing GHG emissions. For example, Exxon Education Foundation's *Exxon Energy Cube*, with videos, books, games and posters, 'implies that fossil fuels in general pose few environmental problems and that alternative energy is unattainable and costly' (Levy and Egan, 1998, p344).

Second, oil companies operating in Europe have been repeatedly exposed to campaigns and boycotts initiated by the green movement. For example, following boycotts of Shell's petrol stations after the Brent Spar incident, it emerged that losses from the boycott could be more costly than the plan to dump the platform at sea (Estrada et al, 1997). It is quite illustrative that the ongoing 'Stop Esso' campaign specifically related to Exxon's climate policy was initiated in Europe and not in the US. Greenpeace-US finds it difficult to raise funding on climate change owing to low public concern. Even though the US ratification of the Kyoto Protocol would have been the key to a viable climate regime, Greenpeace-US restricted its efforts in lobbying for US ratification even before George Bush Jr was elected in the fall of 2000.[33]

Social demand is also important for understanding differences in climate policy. US climate policy is strong on scientific research and weak on action. As noted, the Clinton–Gore administration struggled to develop a viable climate policy in the US, owing to resistance from Congress and the fossil fuel lobby. In April 1993, Clinton announced that the US had committed itself to reducing emissions of GHGs to their 1990 levels by the year 2000. However, after the defeat of the BTU tax, which was seen as the core policy instrument, the fossil fuel industry did not take the stabilization target seriously. The Clinton–Gore administration was forced to shift its focus from taxes to voluntary public programmes. The core of US climate policy over the last decade has comprised a number of genuinely public voluntary programmes aimed at creating markets for energy effective technologies. These programmes do not deal with the oil industry, but are rather general cross-sectoral programmes. ExxonMobil has shown little interest in the 40 different programmes, arguing that none of them have led the company to take action different from what it would have done in their absence.[34]

Compared to the US, the Netherlands, the UK and the EU have set ambitious goals and adopted a wide range of policy instruments; the Netherlands in particular has focused actively on renewable energy. As noted above, the UK and the EU have launched ambitious climate change programmes in which the oil industry participates. In March 2002, EU Member States agreed to be legally bound by the Kyoto Protocol, and the EU ratified the Protocol in May the same year.

For Shell, the Netherlands does not only constitute a 'test country' for pressures and opportunities offered by the public at large. It also represents a test case for what the oil industry can expect from a relatively viable climate policy. The Netherlands has aimed at a 3 per cent reduction in $CO_2$ emissions by 2000 against its 1990 levels. In 2002, the Netherlands was the first of the EU Member States to ratify the Kyoto Protocol. The Dutch use of policy instruments has also been gradually stepped up and comprises taxes, legal regulation and negotiated agreements. A long-term agreement was concluded between the Dutch authorities and the oil industry in 1996. In 1995, the Dutch government decided to increase its share of renewable energy to 10 per cent of total energy consumption by 2020. In 1996, the EU followed up with its intention to double its share of renewable sources of energy by 2010. Combined with regulatory pressure, these initiatives influenced Shell's decision to establish Shell International Renewables in 1997.[35]

The differences between political institutions regulating state–company relationships in the US and most European countries are well documented. These differences are also to some extent reproduced and reinforced by the multi-stakeholder approach of the EU. In the mid-1980s, Vogel (1986) described the US style as the most rigid and rule-oriented to be found in industrial society. Ten years later, Wallace (1995) still held that the US was based on an adversarial, legalistic approach to environmental issues, an approach which has produced an inflexible, fragmented and confused regulatory system. This description of the US approach is shared by the API.[36] The image of the US regulatory system is now changing, but the backbone of US regulatory and compliance models remain essentially intact (Dannenmaier and Cohen, 2000). As a consequence, industry prefers opposing environmental goals to seeking creative solutions. In climate policy, corporate opposition has proved successful and has to a large extent determined the regulatory framework.

The Dutch and UK approaches have been described as almost the opposite. In comparison with the US, Vogel (1986) describes the UK approach as a model where policy makers work closely with the industries. They rely on expert advice from industry and seek industry consent before changing policy, as well as in the implementation phase. This can be illustrated by the appointment of Shell's Chris Fay as a leader for the UK emissions trading scheme. The Netherlands has gone a step further. Partnership with industry based on long-term agreements has been an integrated part of Dutch environmental policy since the first National Environmental Policy Plan (NEPP) in 1989. Shell has a very good relationship with Dutch authorities on environmental matters (Skjærseth and Skodvin, 2001).

In sum, ExxonMobil has operated within a political context that has generated scant new market opportunities for renewables, has exerted little regulatory or societal pressure, and is marked by a lack of cooperation between the state and industry. Of particular importance for the climate strategy of ExxonMobil is the weakness of the state in climate policy and the power of the fossil fuel lobby. ExxonMobil has been convinced that US climate policy would remain weak and that the Kyoto Protocol would never be ratified by the US. According to the company, it has never developed a Plan B in case of US ratification of the Kyoto Protocol.[37] Conversely, Shell and BP have faced a significantly different political context pulling in the opposite direction. The combination of regulatory pressure and market opportunities has led these companies to choose a different climate strategy. These companies have not been able to control the political processes and have thus chosen to exploit new opportunities. In the case of Shell in particular, there is a direct link between public programmes on renewables and the establishment of a fifth core business area on renewables.

## Concluding remarks

Major European and US oil companies in alliance with other industries have affected compliance and compliance systems in different ways. The largest company in the world – ExxonMobil – has played a leading role within the US fossil fuel lobby. The aim of this lobby has been to defeat any mandatory climate commitments at home and abroad, and consequently any compliance systems linked to mandatory obligations. Judged on goal attainment, this lobby has had significant success. US climate policy is still voluntary and the US is no longer a part of the Kyoto Protocol.

One important reason behind this development is the influence of the US fossil fuel industry, organized by the GCC. This coalition represented a significant part of US GDP, and its members utilized a wide range strategies to achieve their goal, including legal threats, media pressure directed at politicians, political donations, and exploiting a number of studies showing the negative economic consequences of the Kyoto Protocol for the US. In addition, ExxonMobil has been very active in disputing IPCC science. These efforts fit well with coercion.

On the other side of the Atlantic, the companies ranked as numbers six and seven in the world by revenue – BP and Shell – have chosen a significantly different strategy. These companies have influenced climate policy instruments and compliance systems by voluntarily setting an example for a proactive corporate climate policy. BP and Shell have taken the lead in adopting and developing GHG emissions targets, reporting and verification routines, and in-house emissions trading systems. As such, these companies have served as models for other companies, as well as for compliance systems in the EU, by means of unilateral action.

The good news from a governance perspective is that the main causes of these significantly different climate strategies are linked to differences in the political context in which the companies operate, rather than in the companies themselves. The combination of cooperation, pressures and opportunities signalled by the EU, the UK and the Netherlands have affected BP's and Shell's anticipation of future climate regulations under the EU and the Kyoto Protocol. Their strategy has been to exploit the market opportunities associated with this scenario by taking the lead in climate policy. Conversely, a conflict-oriented relationship with industry in environmental policy, the lack of regulatory pressures and the lack of new market opportunities in the US has had the opposite effect on Exxon. The success of the US fossil fuel lobby has led ExxonMobil to believe – correctly so far – that mandatory US regulation is unlikely.

The bad news is that social demand, the governmental supply of climate policy and political institutions governing state–company relations in the US are not likely to change significantly in the short term. Given the importance of political context for understanding corporate climate strategy choice, the climate plan presented by Bush in February 2002 could have provided some new incentives. The new plan is, however, voluntary and will apparently not provide sufficient opportunities and pressures to change the climate strategy of companies like ExxonMobil.

The Kyoto Protocol still represents the most potent political force that will affect US multinationals with significant activities in Europe. What consequences might a Kyoto Protocol without the US have for the climate strategies of US companies? This question depends on a number of uncertainties, such as the fate of the Kyoto Protocol itself and the domestic implementation of the Kyoto commitments. Bodansky (2001) argues that the Kyoto Protocol might have both positive and negative consequences for US multinationals in the absence of a viable US climate policy and US participation. Significant negative consequences can weaken the fossil fuel lobby by providing US multinationals with incentives to soften their opposition. US multinationals subject to GHG regulation in Europe cannot take advantage of emissions reduction in the US, since Kyoto does not recognize emissions reductions achieved in non-parties. US multinationals with operations in Europe have to comply with mandatory national standards. This means that ExxonMobil is a party to the Dutch long-term agreement with the oil industry aimed at a 20 per cent improvement in energy efficiency, and Exxon has to pay $CO_2$ taxes wherever they apply.[38] A viable EU climate policy could thus provide US multinationals with incentives to change their strategy, or move GHG emission operations from Annex B countries to developing countries or the US. Moreover, US companies can be exposed to negative public attention and consumer boycotts like the 'StopEsso' campaign. A recent report prepared by a group of dissident investors argues that Exxon's stance on climate policy will hurt shareholder value in the long run. It further makes a comparison between Exxon and the tobacco companies, arguing that the risks involved in Exxon's anti-Kyoto Protocol stance are much higher than the strategies adopted by BP and Shell. Exxon has characterized the report as 'ridiculous'.[39]

Whereas US multinationals may experience negative consequences from the current situation, Kyoto can provide energy-intensive companies in the US with a competitive advantage if energy prices in Annex B countries increase. However, the US exit from the Protocol has had a considerable impact on the demand–supply balance of the carbon market, since the supply of 'hot air' is likely to be greater than the total demand in the carbon market under the Kyoto Protocol. In essence, if the US, which is responsible for about 25 per cent of global GHG emissions, does not re-enter the Kyoto Protocol, the value of carbon is likely to be much lower than anticipated by the US fossil fuel lobby, owing to the potential oversupply of surplus permits (hot air) from Russia and Ukraine. The actual costs, however, remain highly uncertain. US companies may gain a short-term competitive advantage. In the long term, however, incentives for developing new technologies within the world's largest economic market will be limited. In addition, US companies exporting to Kyoto parties may face trade restrictions in the absence of voluntary measures (Jacob, 2001; see, however, Chapter 7 by Stokke).

How the positive and negative consequences add up for US multinationals is still to be seen. However, the European oil giants have already made their marks in the US, and US companies are warming up to Kyoto in Europe. BP and Shell have exerted influence within the GCC, and continue to exert influence within the API and EUROPIA. US multinationals are already subject to different national environmental requirements, but Kyoto will reinforce this problem. There are indications that parts of the US industry may have taken initial steps towards a more proactive stance within the US. The best example of this is perhaps the foundation of the Pew Center on Global Climate Change, a think-tank, in 1998. Shell and BP have taken a leading role.[40] In 2000, the Pew Center comprised 21 major US companies. Today, the number has risen to 37, most of which are included in the Fortune 500. In Europe, large US multinationals like the chemical giant Dupont are preparing for emissions reductions. Roughly half of Dupont's 85,000 employees work outside the US, and Dupont has – in an attempt to anticipate obligations – settled on a plan to reduce emissions by 65 per cent from 1990 levels by 2010. Dupont is the second-largest participant in the UK emissions trading scheme in terms of emissions reduction targets for 2006.

European multinationals operating in the US and US multinationals operating in Europe may further weaken the US fossil fuel lobby. Over time, this may pull towards a convergence between the US and the EU and a higher level of compliance, with the aim of limiting emissions of GHGs as included in the UNFCCC.

## Notes

1 For a more comprehensive study of climate change and the oil industry, see Skjærseth and Skodvin, 2003.

2 'Oil Giant Bids to Replace Climate Expert', guardian.co.uk, 5 April 2002.

3 'UK's Blair Defends London Meeting With Exxon Mobil's Raymond', Bloomberg.co.uk, 24 January 2002.

4 It should be noted, however, that within Europe and North America differences in climate strategies also exist between oil companies. For example, other US-based oil companies such as Texaco-Chevron have chosen a more ambitious approach than ExxonMobil, but they nevertheless oppose the Kyoto Protocol.

5 In 2001, ExxonMobil was the largest company in the world with revenues amounting to US$210,392 million. In 2001, Shell was ranked as number six (US$149,146 million) and BP as number seven (US$148,062 million). www.fortune.com/list/G500/index.html

6 The UNFCCC states that the parties shall adopt policies and measures that would limit man-made GHG emissions and protect and enhance sinks and reservoirs within their territories with the aim of stabilizing GHG emissions at their 1990 levels by the year 2000. The vague character of the commitment makes it questionable whether it can be considered binding and enforceable.

7 In addition, indifferent and innovative strategies are frequently included.

8 See the corresponding discussion of environmental organizations' influence in Chapter 8 by Andresen and Gulbrandsen.

9 These types are borrowed from the literature on leadership. See particularly Underdal, 1991.

10 According to Agrawala and Andresen (1999, p461), the UNFCCC 'was solely a result of US adamancy even in the face of complete isolation'.

11 'Documents Show Energy Officials Met Only With Industry Leaders', *New York Times*, 27 March 2002.

12 'A Dinosaur Still Hunting For Growth', *Financial Times*, 12 March 2002.

13 The question of compliance was barely mentioned, according to the summary record of a workshop organized by the OECD/IEA on industry views on the climate change challenge with special emphasis on the Kyoto mechanisms (BIAC/OECD/IEA, 1999).

14 Personal communication with Glenn Wiser, CIEL, Washington, DC, 20 March 2002.

15 Personal communication with Philip A. Cooney and William O'Keefe, API, Washington, DC, 21 March 2000.

16 'Global Warming Business Group Cools Its Message', Reuters, 9 November 2000.

17 Personal communication with GCC represented by Glenn Kelly and Eric Hold, 2000.

18 Early in 2003, ExxonMobil decided voluntarily to report carbon emissions, and the company is now backing mandatory reporting as a first step towards targets on emissions reduction.

19 'US Senate Marks Earth Day with Global Warming Debate', Planetark, 22 April 2002.

20 'MEPs Reject Mandatory Company Reporting', *ENDS Report*, 26 March 2002.

21 '"Hot Air" Blows Gaping Hole In Emissions Trading Scheme', *ENDS Report*, March 2002.

22 Ibid.

23 It is too early to judge the success of the scheme since trading has taken place only since 2 April 2002. Nevertheless, critics have already questioned the ability of the scheme to deliver 'real' reductions in emissions. Half or more of the reductions are claimed to be not real (hot air) or would have been delivered anyway due to the expected reduction in market shares.

24 Planet Ark, 'EU Parliament Launches Climate Emissions Trading', www.planetark.com/dailynewsstory, 3 July 2003.

25 Personal communication with EUROPIA represented by V. Callaud, 30 November 2000.

26 Personal communication with EUROPIA represented by V. Callaud, 2000.

27 UNICE suggests deleting 'or twice the average market price… [whichever] is the higher', arguing that 'penalties are sufficient to ensure a high degree of compliance, and providing for a penalty twice an average market price, if higher, brings unnessary uncertainty' (UNICE, 2002, p3).

28 The 'beauties and beast' expression is borrowed from Rowlands, 2000.

29 For a comprehensive account of company, domestic and regime explanations, see Skjærseth and Skodvin, 2003.

30 See, for example, Roome, 1992; Steger, 1993; Ghobadian et al, 1998.

31 In 1995, for example, 60 per cent of the population in the Netherlands stated their willingness to pay higher prices for environmentally-friendly products (VROM, 1997).

32 Personal communication with Barend van Engelenburg, Ministry of Housing, Spatial Planning and the Environment, 28 November 2000.

33 Personal communication with Greenpeace, Washington, DC, 23 March 2000.

34 Personal communication with B. P. Flannery and G. Ehling, ExxonMobil, Irving, TX, March 2000.

35 Personal communication with Ir Henk J. van Wouw, Manager of Environmental Affairs, Shell Netherlands BV, 28 November 2000.

36 Personal communication with Philip A. Cooney and William O'Keefe, API, Washington, DC, 21 March 2000.

37 Personal communication with B. P. Flannery and G. Ehling, ExxonMobil, Irving, TX, March 2000.

38 However, a recent consultant report argues that ExxonMobil will actually benefit more than other oil companies from measures such as carbon taxes, since it is less dependent on growth than some of its competitors, like BP. 'Carbon Tax Could Benefit ExxonMobil', *Financial Times*, 7 May 2002.

39 'ExxonMobil Rubbishes Green Investor Report', Planet Ark, 7 May 2002.

40 The objective of the Center is to educate policy makers as well as the general public about the causes and possible consequences of climate change, and to encourage the domestic and international communities to reduce emissions of GHGs.

## References

Agrawala, S. and Andresen, S. (1999) 'Indispensability and Indefensibility? The United States in Climate Treaty Negotiations', *Global Governance*, vol 5, 457–82

Agrawala, S. and Andresen, S. (2001) 'US Climate Policy: Evolution and Future Prospects', *Energy & Environment*, vol 12, nos 2 and 3, 117–37

Betsill, M. M. and Corell, E. (2001) 'NGO Influence in International Environmental Negotiations: A Framework For Analysis', *Global Environmental Politics*, vol 1, no 4, 65–86

BIAC/OECD/IEA (1999) 'Industry View on the Climate Change Challenge with Special Emphasis on the Kyoto Mechanisms', Industry Sector Report, BIAC/OECD/IEA, Paris

Bodansky, D. (2001) *Implications for US Companies of Kyoto's Entry into Force Without the United States*, Pew Center, Washington, DC

Carpenter, C. (2001) 'Business, Green Groups and the Media: The Role of Non-governmental Organizations in the Climate Change Debate', *International Affairs*, vol 77, no 2, 313–28

Christiansen, C. A. and Wettestad, J. (2002) 'The EU as a Frontrunner in Greenhouse Emissions Trading: How Did it Happen and Will the EU Succeed?', working paper, Fridtjof Nansen Institute, Lysaker

COM (2001) 'Proposal for a Directive of the European Parliament and the Council Establishing a Scheme for Greenhouse Gas Emission Allowance Trading Within the Community and Amending Council Directive 96/61/EC', Council of Ministers, Brussels

Cox, R. W. and Jacobson, H. K. (1973) *The Anatomy of Influence: Decision Making in International Organizations*, Yale University Press, New Haven, CT

Dannenmaier, E. and Cohen, I. (2000) *Promoting Meaningful Compliance with Climate Change Commitments*, Pew Center, Washington, DC

DEFRA (2000) *A Summary Guide to the UK Emissions Trading Scheme*, Department for Environment, Food and Rural Affairs, London

ECCP (2001) 'European Climate Change Programme', report, June, http://europa.eu.int/comm/environment/climat/home_enhtm

Estrada, J., Tangen, K. and Bergesen, H. O. (1997) 'Environmental Challenges Confronting the Oil Industry', John Wiley & Sons, Chichester

Gallup, A. and Saad, L. (1997) 'Public Concerned, Not Alarmed About Global Warming', poll releases, 2 December, www.gallup.com/poll/release/pr971202.asp

Ghobadian, A., Viney, H., Liu, J. and James, P. (1998) 'Extending Linear Approaches to Mapping Corporate Environmental Behaviour', *Business Strategy and the Environment*, vol 7, 13–23

Gleckman, H. (1995) 'Transnational Corporations' Strategic Responses to "Sustainable Development"' in Bergesen, H. O. and Thommesen, Ø. (eds) *Green Globe Yearbook*, Fridtjof Nansen Institute/Oxford University Press, Oxford

Hatch, M. T. (1993) 'Domestic Politics and International Negotiations: The Politics of Global Warming in the United States', *Journal of Environment and Development*, vol 2, no 29, 1–39

Jacob, T. (2001) 'US Industry and Climate Change', *Global Change*, vol 7, no 3

Kolk, A. (2001) 'Multinational Enterprises and International Climate Policy' in Arts, B., Noortmann, M. and Reinalda, B. (eds) *Non-State Actors in International Relations*, Ashgate, Aldershot

Levy, D. (1997) 'Business and International Environmental Treaties: Ozone Depletion and Climate Change', *California Management Review*, vol 39, no 3, 54–71

Levy, D. L. and Egan, D. (1998) 'Capital Contests: National and Transnational Channels of Corporate Influence on the Climate Change Negotiations', *Politics and Society*, vol 26, no 3, 337–61

Loreti, C. P., Foster, S. A. and Obbagy, J. E. (2001) *An Overview of Greenhouse Gas Emissions Verification Issues*, Pew Center, Washington, DC

Newell, P. N. (2000) *Climate for Change: Non-state Actors and the Global Politics of the Greenhouse,* Cambridge University Press, Cambridge

OECD (1996) *Environmental Performance Reviews*, OECD, Paris

Raustiala, K. (2001) 'Nonstate Actors in the Global Climate Regime' in Lutherbacher, U. and Sprinz, D. F. (eds) *International Relations and Global Climate Change*, MIT Press, Cambridge, MA

Retallack, S. (2000) 'Economic Globalization and the Environment', *Transnational Association*, vol 4, 181–91

Roome, N. (1992) 'Developing Environmental Strategies', *Business Strategy and the Environment*, vol 1, part 1, 11–25

Rowlands, I. H. (2000) 'Beauty and the Beast? BP's and Exxon's Positions on Global Climate Change', *Environment and Planning C: Government and Policy*, vol 18, no 3, 339–54

Saad, L. and Dunlap, R. E. (2000) 'Americans are Environmentally Friendly, but Issue Not Seen as Urgent Problem', poll releases 17 April, www.gallup.com/poll/release/pr000417.asp, 10 May

Skjærseth, J. B. (1994) 'The Climate Policy of the EU: Too Hot to Handle?', *Journal of Common Market Studies*, vol 32, no 1, 25–45

Skjærseth, J. B. and Skodvin, T. (2001) 'Climate Change and the Oil Industry: Common Problems, Different Strategies', *Global Environmental Politics*, vol 1, no 4, 43–64

Skjærseth, J. B. and Skodvin, T. (2003) *Climate Change and the Oil Industry: Common Problems, Varying Strategies*, Manchester University Press, Manchester

Skodvin, T. and Skjærseth, J. B. (2001) 'Shell Houston, We Have a Climate Problem!', *Global Environmental Change*, vol 11, no 2, 103–6

Skolnikoff, E. B. (1997) *Same Science, Differing Policies: The Saga of Global Climate Change*, MIT Press, Cambidge, MA

Steger, U. (1993) 'The Greening of the Board Room: How German Companies are Dealing with Environmental Issues' in Fischer, K. and Schot, J. (eds) *Environmental Strategies for Industry: International Perspectives on Research Needs and Policy Implications*, Island Press, Washington. DC

Underdal, A. (1991) 'Solving Collective Problems: Notes on Three Modes of Leadership' in *Challenges of a Changing World, Festschrift to Willy Østereng*, Fridtjof Nansen Institute, Lysaker

UNICE (2000) *UNICE Input Ahead of COP 6 at The Hague*, UNICE, Brussels

UNICE (2002) *UNICE Comments on the Proposal for a Framework for EU Emissions Trading*, UNICE, Brussels

Vogel, D. (1986) *National Styles of Regulation: Environmental Policy in Great Britain and the United States*, Cornell University Press, Ithaca, NY, and London

VROM (1997) *The Netherlands' National Communication on Climate Change Policies*, Ministry of Housing, Spatial Planning and the Environment, The Hague

VROM (1999) *The Netherlands' Climate Policy Implementation Plan, Part I: Measures in the Netherlands*, Ministry of Housing, Spatial Planning and the Environment, The Hague

Wallace, D. (1995) *Environmental Policy and Industrial Innovation: Strategies in Europe, the USA and Japan*, Earthscan, London

Watts, P. (1997) 'Taking Action to Earn Trust – Health, Safety and the Environment', speech, 22 September, www.shell.com/library/speech/0,1525,2304,00.html

World Resources Institute (2002) *The Greenhouse Gas Protocol: A Corporate Accounting and Reporting Standard*, World Resources Institute and World Business Council for Sustainable Development, Washington, DC

Chapter 10

# Enhancing Climate Compliance – What are the Lessons to Learn from Environmental Regimes and the EU? [1]

Jørgen Wettestad

## Introduction

The purpose of this chapter is to look at compliance measures used in other international environmental regimes to see if there are any lessons that might be applied in the context of compliance with the Kyoto Protocol. Institutions examined in this context are the Convention on Long-range Transboundary Air Pollution (CLRTAP) regime, the ozone layer regime based on the Vienna Convention and Montreal Protocol, and the internal climate policy of the European Union (EU). There are clearly functional similarities and differences between these three institutions and the climate change regime from which interesting lessons may be learned.

Roughly speaking, each of these three institutional frameworks deals with 'atmospheric' problems (the EU even deals specifically with climate change). Hence, there should be some common challenges related to the measuring and reporting of emissions. With regard to the various North–South compliance challenges, the ozone regime has valuable experience in handling reporting deficiencies and non-compliance within a global environmental policy context. Regarding the handling of the special compliance challenges faced by countries with economies in transition, both the ozone regime and CLRTAP have experience in handling this 'Eastern' dimension. Both these regimes also have several years of experience with the functioning of implementation committees. In the case of ozone, this experience is over a decade long. Finally, in terms of response and enforcement, the EU should offer interesting lessons as a supranational enforcement institution, one that is also starting to address the specific problem of climate change. Although the EU's membership is limited, its countries nevertheless represent a wide range of interests and capacities when it comes to implementation of climate policy. Thus the EU's climate policy can be seen as a microcosm of the Kyoto Protocol, complete with many of the same challenges.

This chapter will describe the evolution of verification, review and response mechanisms within these different institutional contexts, and assess:

1 the major forces that have driven this evolution, including the balance between internal needs and external learning;
2 the extent to which these various mechanisms have actually been used, and their roles in furthering compliance within the regimes in question; and
3 the relevance of lessons learned in these other institutional contexts for the climate regime.

As further elaborated in Chapters 1 and 2 of this book, the Marrakesh Accords adopted in November 2001 put the main finishing touches on the Kyoto Protocol's compliance regime, which, according to the *ENDS Report*, is a 'more far reaching non-compliance procedure than in any existing environmental treaty'.[2] The main elements of the Protocol's climate procedure that this chapter will address, in seeking lessons from other contexts that may be of value, are verification, review and response. In the area of *verification*, national reporting on emissions inventories and other information relating to compliance with commitments is the central ingredient (see primarily Articles 7 and 8 in the Protocol).[3] Here, a central challenge is to ensure on-time and accurate reporting from states with weak administrative capacities.

Second, with regard to *review*, an interesting element of the Protocol's compliance procedure is the establishment of Expert Review Teams, which are to 'provide a thorough and comprehensive technical assessment of all aspects of the implementation by a Party of this Protocol' and report the results back to the Conference of the Parties (COP) (Article 8.3). A key question to be addressed here is whether there is any precedent for this form of arrangement in the three other institutions scrutinized in this chapter. Another striking feature in the institutional structure of the Kyoto Protocol is the establishment of a Compliance Committee, composed of two branches: a Facilitative Branch and an Enforcement Branch. A challenge here will be to ensure smooth communication and cooperation between these two branches so that the Compliance Committee functions effectively. The precise and uncontested identification of non-compliers will be the central compliance challenge within the regime, as experienced observers have noted.[4] Such identification is of course a prerequisite for a meaningful operation of the response system.

Third, in the area of *response*, as can be recalled, Marrakesh negotiators were applauded for agreeing that parties not in compliance with their commitments in the first commitment period, which runs from 2008–12, would have to deduct 1.3 tonnes from their second commitment period assigned amount for every tonne of gas they emitted in excess of their agreed targets in the first commitment period. One critical question to ask in this context is whether there are any precedents for such enforcement procedures under other environmental regimes, and if so, what the experiences with these procedures have been.

Having introduced the climate policy issues around which lessons from the three institutions examined could be drawn, it must be emphasized that there are significant differences and similarities between the ozone layer regime, the CLRTAP regime and the EU climate policy – the three institutions singled out for comparison – and the international climate change regime. Procedures and practice under each regime, with regard to verification, review and response have been shaped both by *the nature of the underlying environmental and cooperative problems at hand* and *the nature of the international political response* emerging to these problems.[5] The following section briefly describes the cooperative contexts of the three institutions and some important ways in which they differ from the global climate change context.

## A cautionary note: Some important differences between the three institutions and the climate change regime

To understand the extent to which it may be possible to apply the lessons learned from the CLRTAP, ozone and EU regimes within the climate change regime, it is important to understand the extent to which the climate regime differs from the institutions to which it is being compared. Overall, it can be noted that in the cases of CLRTAP and ozone, these differences are primarily of a *substantive* nature (in terms of science, politics and the very nature of the issue area), whereas in the case of the EU, the difference is primarily of an *institutional* nature.[6]

The CLRTAP's geographical scope is based in the Economic Commission for Europe (ECE), which means that the regime has not needed to deal with complex and complicated North–South issues.[7] Moreover, for a lengthy period, the commitments adopted within the regime were quite unambitious.[8] Hence, although some countries have experienced troubles with meeting the Convention's volatile organic compounds (VOC) commitments agreed to in 1991, it can be argued that it was not until the adoption of the 1999 Gothenburg Protocol that the CLRTAP regime presented a regulatory challenge comparable to that of the Kyoto Protocol. In addition, and partly related to the relatively unambitious nature of CLRTAP measures, the level of basic mistrust between CLRTAP parties has for some time been low and quite different from the high level of mistrust and multifarious conflict lines that exist within the climate change context. As noted by Harald Dovland, an experienced CLRTAP and climate change negotiator, it is necessary to go back to around 1980 to find a CLRTAP climate comparable to the conflict-ridden climate change context.[9]

Turning to the ozone regime and the Montreal Protocol, compliance challenges, at least for the North, have been far less severe than those faced by the climate regime, as a result of what turned out to be the quite benign character of the ozone depletion problem, and, related to this, relatively easily implementable ozone mitigation policies.[10] Hence, the ozone regime has not been faced with the challenge of handling cases of Northern non-compliance, which will be a central issue in the case of climate change, taking a long-term perspective. Moreover, it can be argued that the richest area of ozone compliance experience, which is with the handling of implementation problems and non-compliance by Eastern countries with economies in transition (CEITs), is not directly transferable to the climate change context due to the different positions of the CEITs within the two contexts. In the ozone context, CEITs have been faced with the burdensome task of phasing out the use of cheap and effective substances, and the lucrative trade in these substances.[11] Somewhat paradoxically, for the CEITs, the climate change context may in fact be much more benign – at least temporarily, and as long as emissions trading is a central policy instrument.[12] This is because industrial restructuring, economic recession and markedly reduced emissions in the 1990s make many of the CEITs net sellers of their assigned amounts under the Kyoto Protocol, which other parties to the Protocol can use to offset their own emissions. Hence the central verification and compliance challenges for the CEITs in the climate context in the short run will be focused on monitoring emissions and trade, rather than on reducing emissions of controlled substances.

The EU's internal climate policy presents a very different institution from the global climate regime. Although the EU is expanding, its membership is basically regional. Moreover, the much broader competence of the EU creates possibilities for far greater interaction with other issue areas in the formulation and implementation of climate policy, as well as greater spillover effects from these policies. In addition, it is important to keep in mind that the EU does not have a climate policy in place comparable, for instance, to the EU ozone regulation.[13] Finally, the dual character of the EU should be noted: the EU is both an *arena* for developing environmental policy and an *actor* in international regimes. Although most attention will here be given to the latter aspect, there are interesting verification and compliance issues related to both dimensions, as well as a tension between these different roles, as will be seen in discussions of the ozone regime.

With these notes of caution for background, let us then address the evolution and functioning of verification mechanisms within the institutions under consideration.

## Reporting and verification: Far from perfect, but improving over time

### *Fundamentally decentralized systems*

The CLRTAP, ozone and EU climate regimes each had fundamentally decentralized reporting and verification systems at the onset. Article 4 of the 1979 LRTAP Convention merely called for information exchange. More specific commitments on emissions reporting followed in subsequent protocols. For instance, the CLRTAP's 1985 Sulphur Dioxide ($SO_2$) Protocol stated that 'each Party shall provide annually to the Executive Body its level of national annual sulphur emissions, and the basis upon which they have been calculated' (Article 4). In addition, Article 6 called for information on national programmes, policies and strategies, providing that: '[t]he Parties shall ... develop without undue delay national programmes, policies and strategies which shall serve as a means of reducing sulphur emissions or their transboundary fluxes, by at least 30 per cent as soon as possible and at the latest by 1993, and shall report thereon as well as on progress towards achieving the goal to the Executive Body'. A central element in the monitoring of state performance under the CLRTAP has been the parties' annual reporting to the Secretariat on emissions, and on the procedures and measures parties have adopted for the abatement of emissions and the measurement of acid precipitation.

Within the ozone layer context, the framework 1985 Vienna Convention – which lacked any specific regulatory components – only contained a general paragraph on the transmission of information 'in such form and at such intervals as the meetings of the parties to the relevant instruments may determine' (Article 6). In line with the regulatory development and specification which took place in the 1987 Montreal Protocol, a more specific compliance regime began to take form. In Article 7, the parties were required to provide annual data on production, imports and exports of all controlled substances. Hence, the data reporting component of the ozone regime is based on two main pillars.[14] First, reports from the parties are required on the production, export and destruction of ozone-depleting substances, providing 1986 baseline data and annual data thereafter.[15] Second, the Secretariat receives these reports, processes and analyses them, and produces summary compliance reports (which are public) to the Implementation Committee (further described below) as well as to the Meetings of the Parties.

The importance of industrial and technological substance substitution and related trade issues have raised concerns about industrial secrecy and competitiveness in the ozone context. Hence, there has been a confidentiality dimension surrounding the required reporting which has added to the decentralized character of the system (Parson and Greene, 1995, pp37–8). Nearly all data involved are confidential, and only national totals of production and consumption of regulated substances are reported publicly. Due to commercial considerations, only broad groups of chemicals are reported, and exports and imports are not reported as separate categories. Moreover, the refusal of the EU countries to provide consumption data at the national level makes it impossible to assess the compliance of individual EU Member States; EU members report only to the EU Commission, and their reports remain confidential. As indicated earlier, this example illustrates the tension that sometimes exists between the twin roles of the EU as a policy arena and as an external actor in international regimes.

The EU climate regime also began with a decentralized reporting and verification system. The complicated and slow initial development of EU climate policy is well summed up elsewhere (e.g. Skjærseth, 1993; Haigh, 1996; Wettestad, 2000). The main elements of the initial climate policy 'package' consisted of a proposal for a carbon dioxide ($CO_2$) tax, an energy efficiency framework directive (SAVE – Specific Actions for Vigorous Energy Efficiency), a decision on renewable energies (ALTENER), and a monitoring mechanism. The decision to establish an EU monitoring mechanism for $CO_2$ and other greenhouse gas (GHG) emissions was finalized in June 1993, and required each Member State to 'devise, publish and implement national programmes for limiting their anthropogenic emissions of $CO_2$', with the objectives of contributing to $CO_2$ emissions stabilization in the EU and fulfilment of the Union's commitments under the 1992 Framework Convention on Climate Change (FCCC).[16] National programmes, which were to be updated periodically, were also to include information on 1990 $CO_2$ emissions, inventories of emission sources and sinks, details of national policies and measures that contribute to limiting emissions, trajectories for national $CO_2$ emissions between 1994 and 2000, and measures being taken or envisaged for the implementation of relevant EC legislation. Member States were to regularly provide the Commission with information on emissions inventories and progress with regard to the national programmes, and the

Commission was then to report to the European Parliament and the Council. The 1993 Decision also established a specific EU Monitoring Mechanism Committee.[17]

## *Some additional, centralized features*

Although the overall picture is one of decentralized reporting and verification systems, there are also certain centralized features that are notable. Under the CLRTAP regime, for example, parties can, to some extent, draw upon additional physical evidence to obtain independent verification of reported emissions. This capacity stems from the EMEP monitoring system (Cooperative Programme for Monitoring and Evaluation of Long-Range Transmission of Air Pollutants in Europe) which grew out of an Organization for Economic Cooperation and Development (OECD) air pollution study programme established in 1972. The main objective of EMEP has been to provide governments with information on the deposition and concentration of pollutants, along with data on the quantity and significance of the long-range transmission of pollutants. The programme has three main elements: emissions data, measurements of air and precipitation quality, and atmospheric dispersion models. The EMEP sampling network consists of some 100 stations in 33 countries, with work coordinated by three international centres: two in Oslo and one in Moscow. A specific EMEP financing protocol was established in 1984. Article 5 of the 1985 Sulphur Protocol, which addresses calculations of transboundary fluxes, provides that 'EMEP shall in good time before the annual meetings of the Executive Body [EB] provide to the EB calculations of sulphur budgets and also of transboundary fluxes and depositions of sulphur compounds for each previous year within the geographical scope of EMEP, utilizing appropriate models.'

The EMEP system led Sand (1990, p259) to maintain: 'Few other international agreements can be said to come equipped with verification instruments of this calibre.' However, it is important to acknowledge that the EMEP reports are also partly based on national reports,[18] some countries have failed to report emissions, and monitoring coverage in Southern Europe has been pointed out by EMEP itself as insufficient (Levy, 1993, p89). In addition, it has been called a paradox that the greatest challenge to producing useful compilations of data from national reports has not been low technical capacity, but resistance to harmonization of reporting methodologies from states with highly developed national systems (Wiser, 1999, p6). Overall, the impression is that although the EMEP system in theory does provide additional verification capacity, its practical effects have been very moderate.[19]

Within the ozone layer context, evidence on the state of the ozone layer produced by the Global Ozone Observing System (GOOS) also provides some additional physical evidence that Montreal Protocol parties can potentially draw upon for external verification of progress in achieving Convention goals. The system was established by the World Meteorological Organization (WMO), under the umbrella of the United Nations Environment Programme (UNEP)'s Global Environment Monitoring System (GEMS), and has approximately 140 monitoring stations worldwide, which are supplemented by remote sensing by satellites. It is capable of providing data on both the horizontal and vertical distribution of ozone, as well as on the total atmospheric concentration of ozone.[20] Hence, it provides a form of verification by remote sensing which requires no access to the territories of Montreal Protocol parties. While, on the one hand, the system produces only a very rough and indirect check of the information produced by the parties themselves, on the other, its results *have* actually been referred to as further evidence of the cuts in emissions reported by the parties. According to Reuters, at the 1999 Meeting of the Parties in Beijing, a chief officer of the UNEP Montreal Secretariat declared: 'scientists have reported that some potent ozone-depleting substances (ODSs) at the upper atmosphere have decreased and this is a very positive sign... *This is the result of the elimination of ODSs from countries such as the United States, Japan, Canada, Western Europe and some developing countries*'(my emphasis).[21]

In the EU environmental policy context, in addition to Member State reporting, there is an informal procedure through which the Commission records and examines complaints lodged by citizens, firms, non-governmental organizations (NGOs) and national administrations.[22] As noted by Tallberg (2002, p616): 'The complaint procedure offers a form of monitoring that is more resource efficient than systematic in-house inquiries, provides access to information otherwise unobtainable, and points to areas of EU legislation that may be particularly ambiguous and in need of clarification'. This informal channel is of course also of interest and importance in the more specific EU climate change context.

## *Improvements over time, but remaining deficiencies*

Initial reporting must be characterized as deficient in all three institutions, with certain variations. Nevertheless, there has been a clear trend of improving reporting over time, though some problems and deficiencies still remain.

With respect to CLRTAP reporting, changes in the East–West relationship from the early 1990s on have contributed greatly to increased Eastern transparency and more reliable data. However, according to experienced participants in the regime, the weakest point in the present scientific political complex is still the emission data.[23] As pointed out by di Primio (1996, p43), 'some data can be questioned because time series are overtly incomplete, and/or show constant rounded figures for long periods, and/or consider as provisory emissions for previous emissions for previous years'.

A central factor which has contributed to an improved reporting system in the CLRTAP is the establishment of a specific Implementation Committee. As will be further elaborated below, this Committee was established in connection with the 1994 second sulphur protocol. At its first meeting, which took place in 1998, the Implementation Committee concluded that 'very few Parties have been respecting the reporting deadlines set by the Executive Body'.[24] The Committee further noted that information on the methodologies used by parties to calculate their emissions was not systematically reported or recorded, [25] and improving this aspect of the CLRTAP compliance system became a central focus for the Implementation Committee. Much time has since been spent on improving reporting questionnaires and linking them more directly to the protocol commitments.[26] In its fourth report, the Implementation Committee notes that the completeness of reported emission data has improved, in some cases 'significantly'.[27]

The ozone regime also experienced substantial initial reporting problems. Many countries did not report on time. For instance, of the 136 parties due to report for 1994, only 57 parties (42 per cent) had reported by October 1995.[28] Moreover, many reports were 'consistently incomplete'; of the 126 parties due to report for 1993, only 69 parties (54 per cent) had reported complete data by October 1995.[29] Baseline data (for 1986 or 1989) were also missing for a number of countries. Not surprisingly, as reporting capacity was higher in the industrialized country parties than in developing country parties (also referred to as the Article 5 countries under the Montreal Protocol) or CEITs, reporting practice was also better in industrialized countries than in Article 5 countries or CEITs (Oberthür, 1997, p11). However, according to ozone negotiators, by the mid-1990s reporting could already be regarded as surprisingly good overall, given the global participation in the regime and hence substantial variations in administrative capacities. The will to report was clearly present.[30]

Although reporting in the ozone regime has improved somewhat over time, data quality still 'remains moderate at best' (Oberthür, 2001, p96). The picture is 'sufficient' with regard to the major ODSs (i.e., chlorofluorocarbons (CFCs) and halons). This is to a large extent due to a fine job done by the Secretariat, with an elaboration of data forms and accompanying instructions for filling these forms out. However, 'considerable problems persist, as is obvious from discrepancies which exist between global production and consumption data as well as between production data available from industry and production reported to the Ozone Secretariat' (Oberthür, 2001, p96). Still further discrepancies exist between data reported to the Ozone Secretariat and those received by the Multilateral Fund Secretariat (Oberthür, 2001, p96). The Multilateral Fund was established in 1990 to support phase-out measures in developing countries, financed by developed countries, and has its own Secretariat in Montreal. At the October 2001 meeting of the Ozone Implementation Committee, it was concluded that more than 50 Article 5 parties (i.e., developing countries) had failed to submit data for 1999, and 57 Article 5 parties and 19 non-Article 5 parties had not reported data for 2000.[31]

The EU GHG monitoring system also experienced substantial initial problems, with patchy reporting at the outset. The first evaluation was on time (March 1994), but incomplete. The second evaluation was delayed for one year and finally completed in May 1996. Although reports had improved in quality, information was still insufficient to evaluate progress in a satisfactory way. For example, reporting was insufficient from central member states such as Germany, France, Italy and the UK. Although reports should have been published both in 1997 and 1998, none appeared, apparently due to the combination of complicated tasks (e.g. non-comparable data) and strained Commission capacities. In April 1999, the monitoring mechanism was revised and strengthened in order to track progress with the much tougher and more specific Kyoto targets.[32] In general, EU Member States are to determine their GHG emissions in accordance with the methodologies agreed upon in Kyoto and report data on emissions, sinks, policies and measures to the Commission on an annual basis. Nevertheless, in April 2002, it was reported that the Commission had sent warning letters to seven countries for failing to provide GHG emissions and sinks data for the year 2000.[33]

An important recent development in EU climate policy is the adoption of a directive establishing a system for EU-wide emissions trading, which was put forward by the Commission in October 2001 and finally adopted in October 2003.[34] The EU directive contains guidelines for monitoring and reporting of emissions (Article 14 and its Annex IV).[35] Because emissions trading is a key flexibility mechanism under the Kyoto Protocol, the design and experiences gained with the EU's own internal verification and compliance system for emissions trading will be of great interest in the evolution of a global system under the Kyoto Protocol.

## Review systems: COPs with more teeth over time

### *The initial systems: Mild and easy*

The main compliance review within CLRTAP takes place at the yearly meetings of the Executive Body. The Secretariat plays an important role in this connection, by preparing annual reviews and four-year major reviews. The latter serve as a more comprehensive review of national abatement strategies and policies. The most recent four-year review was carried out in 1998.[36] At least up to the mid-1990s, the potential inherent in this review process was not fully utilized. Serious compliance discussions were non-existent, as there were no provisions in the Convention or its protocols until quite recently that authorized the Executive Body to examine critically the data provided in national reports (Szell, 1995). At most, countries made oral clarifying statements at Executive Body meetings.[37] The lack of emphasis on compliance must of course be seen partly in light of the East–West non-confrontational dimension prevalent in the regime. Does this then mean that the verification instruments, even of this calibre, were without importance for state policies? The impression is that the very existence of EMEP data contributed somewhat to reporting accuracy in the West and hence general confidence building within this group of countries, but not as a very important aspect of the process.

Within the ozone regime, in a very interesting institutional innovation, an Implementation Committee was established in 1990.[38] A specific non-compliance procedure followed in 1992, which emphasized both a non-confrontational method of operation and decision making firmly left to the Meeting of the Parties.[39] The committee was composed of ten parties, two from each of the five main geographical regions of the world, elected for a two-year period. The committee's main formal functions were to consider and report on any complaint from, or reference by, a party; to consider and report on the annual report of the Secretariat; to request, where necessary, further information on cases before it; to undertake, 'upon the invitation of the Party concerned', data compilation inside the territory of that party; and to exchange information with the financial mechanism of the Montreal Protocol.[40] Moreover, the committee could decide to present cases of potential non-compliance to the Meeting of the Parties. Formally, the outcome of this process could be trade sanctions against states unwilling to sign the agreement or abide by its provisions. Initially, the Committee's work mainly focused on reporting problems, especially related to baseline data, in close cooperation with the Secretariat. 'Mild' forms of pressure and technical assistance from the committee contributed to improving reporting (Parson and Greene, 1995; Victor, 1998).

Among the three institutional frameworks examined in this chapter, it is in the review process that the different nature of the EU's internal climate regime becomes the most pronounced. The EU's 1993 decision establishing a monitoring mechanism for GHG emissions also established a specific EU Monitoring Mechanism Committee. This committee was set up as a 'regulatory' committee, giving Member States an important role in relation to the Commission.[41] However, the formal review procedure was a quite standard EU one: Member States were to provide the Commission with information on their emissions inventories and progress on national programmes on a regular basis; the Commission was then to report annually to the European Parliament and the Council, assessing the extent to which progress in the Community was sufficient to fulfil established climate policy objectives. As indicated in the section above on reporting, the monitoring mechanism's early years were not very impressive. The second round of review, which took place in 1996, was quite critical in tone. Among other things, the review pointed out that it was not possible to produce a Community emissions trajectory for the year 2000 due to differences in assumptions and methodologies used by Member States. Germany, France, Italy and the UK were explicitly criticized for inadequate reporting on the implementation of measures (Hyvarinen, 1999, p193).

### More teeth over time

Within all three institutions examined in this context, more strict review procedures and practices have evolved over time. This is not least so within the CLRTAP. In an important formal development, the 1991 VOC Protocol required the parties to establish a mechanism for monitoring compliance with the Protocol.[42] This call was followed up and made more specific in the CLRTAP's 1994 second Sulphur Protocol, when Article 7 formally established a specific Implementation Committee, mandated to review implementation and compliance, including decisions on 'action to bring about full compliance with the protocol'.[43] The more specific and practical establishment of the Implementation Committee took place in 1997,[44] when the Committee's mandate was broadened to cover the review of compliance with all the CLRTAP protocols. Within international environmental politics in general, the Implementation Committee established within the ozone layer regime has served as an institutional model for other regimes, including CLRTAP.[45] This is a clear example of inter-institutional learning. Apart from 'institutional diffusion' from the ozone regime, important background factors for tougher compliance procedures in the CLRTAP include the fundamental changes in the East–West relationship and greater openness in the East. These changes have provided a much more beneficial setting for critical follow-up discussions, both at the international and national level, than was the case in the 1970s and 1980s. In addition, increasing regulatory sophistication in the CLRTAP regime has created an increased need for improved institutional procedures.

The CLRTAP Implementation Committee is composed of eight legal experts from the parties. Experienced negotiators have emphasized the value of having both technical and legal expertise represented on the committee.[46] At its first meeting, the Committee decided that it would take all decisions by consensus and any report to the Executive Body on a specific party would first be shown to 'and if necessary discussed with' that party.[47] The important role of the Secretariat was also indicated, and given the committee's limited resources, 'the function of the Committee was to reach its conclusions on the basis of analyses carried out by the secretariat or experts'.[48] Initially, the Implementation Committee concentrated on reviewing reporting procedures and practices. Part of this work consisted of publishing overview tables of reporting 'scores'. A main part of the Committee's work, and hence what may be characterized as its 'compliance strategy', has been to increase transparency. The committee has adopted a strategy of gradually increasing pressure on parties that have not complied with reporting obligations. On the first occasion, non-compliance is noted without highlighting the relevant party's name; if it happens again, the party's name is revealed; and the third time, the committee includes the party in a recommendation to the Executive Body urging it to take action to achieve reporting compliance.[49] The committee also systematically reviews compliance with the various protocols.

Within the ozone regime, it appears that the review process and the work of the Implementation Committee also have developed sharper teeth over time. Still, only limited progress has been made. As noted by Oberthür (2001, p22):

> Despite the annual reports of the Secretariat, which have enabled Parties as well as interested groups to check the data presented and ask for reviews or corrections, the existing data have continued to elicit limited reaction from both Parties and others... The lack of any review function built into the system is apparently not made up for by review activities of Parties or others, including non-governmental organizations.

The EU's climate policy monitoring mechanism was also revised and strengthened in April 1999 (as Decision 1999/296), partly in response to the tougher and more specific commitments adopted in the 1997 Kyoto Protocol. The mechanism's initial focus on $CO_2$ was broadened to all GHGs not covered by the Montreal Protocol. Although the mechanism contains much of the same structure as introduced in 1993, the Commission is now required to take further steps to promote comparability and transparency of national inventories and reporting (Article 3.2). Moreover, the support of the European Environment Agency is now explicitly emphasized (Article 5.4). The first progress report under Decision 1999/296 indicates a certain improvement in the review capacity of the Commission, as reporting has improved a bit (Haigh, 2002, pp3–5). To what extent have these experiences influenced the review system outlined in the newly-established EU emissions trading system? Article 15 of the EU Directive provides: 'Member States shall ensure that the reports submitted by operators pursuant to Article 14(3) are verified in accordance with the criteria set out in Annex V, and that the competent authority is informed thereof.' Thus the main responsibility for review is left at the Member State level, likely due to the existing subnational nature of the regulation of individual operators.

## Response: Much facilitation and little enforcement?

There are some basic similarities between the institutions examined with respect to the triggering of non-compliance procedures and the initial steps of the response process. First, within both the CLRTAP and the Montreal Protocol regimes, procedures can be triggered by submissions by non-complying parties themselves, as well as by other parties and the Secretariat. In both contexts, an increasing number of non-compliance cases have been seen in recent years. Not surprisingly, the handling of non-compliance was first experienced in the ozone context. The non-compliance procedure was invoked for the first time in 1995, with the looming non-compliance of five CEITs (Belarus, Bulgaria, Poland, Russia and Ukraine), who invoked the procedure themselves (Victor, 1998, p155). The 'CEIT problem' continued in the years that followed. In 1998, for example, the Implementation Committee discussed the non-compliance of nine parties: Azerbaijan, Belarus, Czech Republic, Estonia, Latvia, Lithuania, Russian Federation, Ukraine and Uzbekistan.[50] This time, the non-compliance procedure was triggered by the Secretariat. At the 2001 MOP, the Implementation Committee concluded that 24 parties appeared to be in various stages of non-compliance – most with respect to increasing CFC consumption.[51] Of these 24 non-compliant parties, 16 were Article 5 parties.

The overall nature of the response taken by the ozone regime has clearly been one of facilitation. As a general basis for response within this regime, an 'indicative list' of measures was produced by the MOP in 1992. This list included rendering 'appropriate assistance', including technical and financial assistance, and assistance in the collection and reporting of data; issuing cautions; and the suspension of treaty rights (Wiser, 1999, p28). In practice, the regime has leaned on a combination of the first two of these three basic options. The approach taken within the Implementation Committee to the CEIT cases in the mid-1990s emphasized various forms of assistance to secure compliance – *without* offering extensions or adjustments of the regulatory obligations undertaken.[52] This established a course of action (Greene, 1996) and approach that has continued.[53] For instance, when the 1998 MOP reviewed the non-compliance of eight parties, the parties recommended that the Global Environment Facility continue to assist the non-compliers, 'while cautioning them that stricter measures [would] be imposed if they [did] not adhere to their new benchmarks for phase-out'.[54] Russia has been given special attention and facilitated assistance. Prior to the 1998 MOP, ten donors pledged a special contribution of US$19 million to help shut down Russian CFC and halon production facilities.[55] This led Russia, at the 2001 MOP, to warmly thank various donor countries and the World Bank for assistance in phasing out ODS production.[56]

However, these various forms of assistance have been accompanied by cautioning calls for clarifications and performance improvements, using transparency to put a certain subtle pressure on the parties. In fact, at the MOP in 2001, a quite sharp tone can be noted: 'For the Parties with economies in transition whose non-compliance with the Protocol has previously been determined by the Parties, the Committee requested the Secretariat to send strongly worded letters alerting them to their continuing non-compliance and requesting [additional] information'.[57] However, a suggestion at the same meeting, from a group led by the US, that Multilateral Fund assistance might in the future be withheld from non-compliers, was rejected.[58] According to Norwegian negotiators, given the established and tight relationship between assistance from the Fund and phase-out activities in the developing countries, this issue is a very complex and sensitive one.[59]

Within the CLRTAP, the explicit and specific discussion of countries' non-compliance with protocols is of a more recent nature. In 2000 the Implementation Committee for the first time considered a case of possible non-compliance, as Slovenia warned about its potential inability to comply with the 1994 Sulphur Protocol.[60] This was followed in 2001 by several cases of non-compliance with the 1991 VOC Protocol. Norway, Finland and Italy came forward with submissions about their non-compliance.[61] The CLRTAP's response to cases of non-compliance must clearly be characterized as a 'managerial' one. For instance, in the case of Slovenia, the Executive Body's response includes an invitation to other parties 'to examine ways in which they could assist Slovenia in reducing emissions from the respective plant, for instance through the provision of equipment'.[62] In the case of Norway, which had failed to meet the VOC Protocol target, the Implementing Committee 'expresse[d] its concern at the failure' and 'urge[d] Norway to fulfil its obligations ... as soon as possible'.[63]

The EU stands out in principle as the most clear enforcement institution of the three examined in this context of response. This is primarily related to the possibility within the EU system of proceedings before the European Court of Justice (ECJ) and the availability of fines for non-compliance. Although 90 per cent of cases of potential non-compliance are settled by communication and 'compliance bargaining' between the Member States and the Commission, there has been an increasing tendency in the recent years of cases going all the way to the ECJ. Still, in the field of environmental politics, there is only one example so far of a state actually being fined.[64] In the case of the climate change monitoring mechanism, the Commission has adopted a clear managerial approach. For example, when the second evaluation of reporting revealed a number of shortcomings, the Commission responded by considering a specific workshop on the production of emission trajectories, and the Monitoring Mechanism Committee envisaged a working group to provide assistance (Hyvarinen, 1999, p193).

However, not surprisingly, in the EU emission trading context, a much more enforcement-oriented approach is in the pipeline. In the EU's emissions trading directive, the EU has set a financial penalty at a rate of €40 per excess tonne of $CO_2$ emitted in the pre-Kyoto period, and a penalty of €100 during the Kyoto Protocol's first commitment period. The experience and success of the US sulphur trading scheme may have played a key part in the inclusion of financial penalties in the EU emissions trading scheme, as the US programme's excellent compliance record was acknowledged in the draft directive.[65] When the draft directive was put forward, the NGO community warmly welcomed the possible inclusion of a 'strong, clear compliance system', arguing that without such a system 'companies will not invest in emission cuts or seek to buy allowances' (Climate Network Europe, 2002).

## 'The essential eight': Lessons to be learned in the case of climate change

Let us then sum up the main lessons to be learnt from the three institutions under scrutiny in this context, bearing in mind the cautionary aspects noted in the introduction. Turning first to the issue of reporting and verification, eight central lessons stand out.

### *An institutional warm-up period is inevitable*

All three institutions experienced an initial institutional warm-up period for reporting, with a significant number of delayed, incomplete or even missing reports. Deficiencies were seen at both the international and national levels, with unclear reporting procedures produced by Convention secretariats and a lack of both will and capacity to report seen at the domestic level. In the case of ozone, which is most similar to the international climate regime in institutional structure and scope, some might argue that this warm-up period is still not complete after almost 15 years, as reporting continues to be problematic. Experienced observers point out that, apart from the cases of CFCs and halons, considerable problems persist. Moreover, it is somewhat surprising to note that the EU's internal climate regime, which is comparatively strong as an institution, has experienced a lacklustre and slow start to its monitoring mechanism. While the mechanism was reorganized in 1999, as noted above, recent reports of missing data from a number of EU countries make it unclear whether reporting really has improved significantly in response to this initiative.

The experiences of these three international environmental regimes are to some degree relevant also in the case of the international climate change regime. Because reporting under the UNFCCC has been carried out for a decade already, it may be expected that some of these initial warm-up problems have already been addressed. Moreover, climate reporting is considerably supported and enhanced by the activities of the international review teams. On the other hand, as further explored in Chapter 4 by Berntsen and associates, it is clear that measurement and reporting challenges are in several ways more severe in the Kyoto Protocol context than they have been in the ozone/Montreal context.

### *Wise institutional engineering can speed up improvements*

In all three cases scrutinized here, reporting has improved considerably over time, largely due to the improvements international institutions have made in their procedures and practices. In both the ozone and CLRTAP regimes, secretariats have played an important role in bringing about procedural improvements – in both cases in close cooperation with the relevant Convention's implementation committees. From 1998 on, the CLRTAP's Implementation Committee has devoted a considerable amount of time to the improvement of reporting questionnaires. As a result, in recent years reporting has in some cases improved significantly. In the ozone context, a similar effort was conducted by the Secretariat from the mid-1990s on. Although problems still persist, things have improved considerably, in particular with respect to the major ODSs. Scientific/technical regimes bodies also have been included in this work. The EU's 1999 reorganization of the EU Monitoring Mechanism also improved methodologies and procedures, although the results of these procedural improvements are uncertain so far.

As noted above, these experiences indicate that the climate regime may also require a warm-up period, in which procedures and institutional solutions are tested out and fine-tuned to the particular international policy problem at hand. While the climate change challenge differs significantly in both intellectual and political terms from the problems of transboundary air pollution and ozone depletion, there are surely some common features in the procedural improvements carried out by the three other secretariats and implementation or compliance committees that are of relevance also in the climate change context. Hence, awareness of these common features and relevant experiences may enable the climate regime to skip some stages, design better procedures than the three institutions scrutinized here, and start the institutional fine-tuning process more quickly and effectively.

### *Independent inputs should be encouraged, but verification expectations should be kept moderate*

Although all three institutions first and foremost rely on national reporting within a fundamentally decentralized context, various independent data (networks) exist in all cases. The most prominent example is probably the EMEP monitoring system in the context of transboundary air pollution, which offers a certain possibility for independent checks of national reports. In the case of ozone, the GOOS provides data on the state of the ozone layer. This offers, of course, only a very indirect check on what the parties say they have done. And, in the EU, some independent evidence is produced through an informal complaint procedure. However, as further discussed below, the interest and willingness to use this data actively for verification purposes and checks on national reporting seem moderate in all three institutions.

In the climate change context, there are first and foremost the Expert Review Teams, which serve as checks on the governments' reports. As no such review teams exist in the three institutions focused on here, this must be noted as one (of several) areas where the climate regime is an innovator and forerunner in international environmental politics.[66]

### *There is a trend towards greater transparency: Climate crooks will be identified!*

Within all three institutions scrutinized, there is a clear trend towards tougher review procedures in the form of much more explicit naming (and shaming) of laggards and non-compliers and tougher language more generally. As noted above, within CLRTAP, an interesting three-stage procedure has been developed, with the explicitness of this naming gradually stepped up if non-compliance persists. This development can of course partly be seen as a result of the parties over time developing a foundation of familiarity and trust, allowing tougher language to be used without threatening the very cooperative fundament itself. Similar explicitness and toughness within the climate regime must develop gradually over time and cannot be expected for some time yet. However, it can also be argued that the trend witnessed in the three institutions is part of a more general trend in international and national politics towards greater transparency and a 'right to know', as perhaps evidenced by the adoption of the 1998 Aarhus Convention on Access to Information, Public Participation in Decision-making and Access to Justice in Environmental Matters, and a related directive within the EU context. Thus the climate regime starts at a time when there is a generally higher level of transparency in environmental matters, and the need for a warm-up phase in this regard may be less pertinent.

But will non-compliers really be identified in a precise and uncontested manner? As indicated in the introduction, this is one of the key climate compliance challenges. Again, the international climate regime's Expert Review Teams represent an institutional advantage, though a more elaborate answer to the question will partly depend upon more technical issues with regard to the measuring of sources and sinks – issues which are explored in more detail in other chapters in this book. However, evidence from the three scrutinized institutions indicates that such identification may very well often take place on the basis of self-reporting of non-compliance, reducing the need for complicated discussions and decisions by regime bodies on precise and uncontested non-compliance. In the case of CLRTAP and VOCs, for example, all three cases of non-compliance in 2001 arose through self-reporting. The lesson that might be drawn is that self-reporting may put the non-compliant party in a better light and position than it would be in if it were to wait for the regime bodies to act on its potential non-compliance.

### *Maintain a close connection between the Facilitative and Enforcement Branch of the Compliance Committee*

As already indicated, implementation committees have been very important institutions in both the CLRTAP and ozone regimes, and have made the review process more meaningful. The CLRTAP Implementation Committee initiated both a process to improve reporting procedures and a systematic, more thorough review of compliance with the multitude of protocols adopted over time. The ozone Implementation Committee has facilitated procedural improvements in cooperation with the Secretariat, and has provided assistance in the reporting process.

The general importance of an implementation committee has clearly been noted by the climate change negotiators. The more specific two-branch approach adopted in the climate context can probably partly be seen as a response to the complex technical and political terrain experienced and envisaged in this context. By separating facilitation and enforcement, the complex challenges are broken down into hopefully more manageable pieces. But are there then organizational insights from the three institutions which may shed light on the prospects for this two-branch approach? The CLRTAP may offer some relevant lessons. As indicated earlier, Harald Dovland, who has participated in the Implementation Committee, emphasizes the value of having both technical and legal expertise represented there – implying the need for easy communication and a high level of communication between the Facilitative and Enforcement Branches of the climate change Compliance Committee.

## *Ways to engage civil society should be explored*

The EU context provides interesting experiences in the engagement of civil society in the process of identifying laggards and non-compliers. Its informal complaint procedure offers a resource-efficient monitoring system which provides access to information otherwise unobtainable and helps to highlight ambiguous legislation. Probably as a reflection of the differences between an international regime and the supranational EU, there are no similar procedures and practices in the two other international regimes scrutinized in this chapter, so expectations should be moderate also in the case of climate change. However, as further explored in Chapter 8 by Andresen and Gulbrandsen, there are opportunities for NGOs to participate in compliance proceedings under the climate regime, and NGOs can function as watchdogs in connection with projects under the Kyoto Protocol's flexibility mechanisms. Overall, the general point seems valid: in a complex international setting, regime bodies and state authorities cannot keep sight of all relevant activities, and should in principle welcome the additional information and critical perspectives provided by civil society organizations and individuals.

## *Assistance and compliance facilitation will be the central response challenge*

With regard to responses to cases of non-compliance, there is a clear trend in the three institutions toward facilitation, which takes place in a variety of ways. In the case of Slovenia's $SO_2$ problems under the CLRTAP, other parties were asked to consider ways to assist Slovenia in achieving compliance, including through the provision of equipment. In the CEITs' difficulties under the ozone regime, the response has also been facilitative, mainly consisting of financial assistance from the GEF and groups of other parties. In the case of the EU and the monitoring mechanism, inadequate compliance was met with a workshop on the production of emission trajectories. Even in the more general cases of non-compliance with EU environmental directives, communication and 'compliance bargaining' sort out 90 per cent of the cases.

In the case of the international climate change regime, it is the inclusion of sanctions that has attracted the most attention – and praise! As commented upon below, this element of the compliance system is undoubtedly important as a hidden stick, but it is highly debatable whether climate change really is radically different from other international environmental institutions in which various forms of management have been the central type of response. This includes the EU's internal climate policy, which in comparison has both fewer and more homogenous parties than the climate regime and has been equipped with formal sanctioning instruments that are likely stronger than those available in the international climate case. It can also be argued that there is little use in setting stricter reduction targets for a subsequent commitment period if a significant reason for non-compliance is the incapacity to follow up previous commitments. Such cases require various forms of assistance, and responding only in the form of sanctions may only make matters worse.

### *The possibility of enforcement will still be important as a 'hidden stick'*

Emphasizing the importance of management does not rule out the importance also of having a credible 'hidden stick'. Such a credible and not-so-hidden stick has clearly been important in the EU context. The fact that some cases go all the way to the ECJ, and countries have started to be fined for environmental negligence, acts as a sobering backdrop to the implementation processes. In the emissions trading context, a financial penalty is envisaged. The EU looks set to establish a trading system by 2005 and valuable compliance lessons may be learned for the global regime to build upon. Also in the case of ozone, a hidden stick in the form of suspension of treaty rights has formally been in operation. According to Wiser (1999, p36), it is the Montreal Protocol's linking of carrots (in the form of technical assistance) and sticks (in the form of threats to cut down or terminate this assistance) which has functioned effectively.[67] A recent call to make greater practical use of these sticks has been made, but has so far been rejected.

As noted, the climate change negotiators have received much praise for the system agreed upon in Marrakesh, with a central aspect being the requirement that non-compliers will have to deduct 1.3 tonnes from their second commitment period assigned amounts for every tonne of gas emitted in excess of their targets in the first period. Given the fact that the economic and political stakes are much higher in the case of climate change than in most other international environmental regimes, particularly from a long-term perspective,[68], the praise heaped upon the enforcement regime is probably well deserved. But if there is a lesson to be learned from the three institutions scrutinized in this chapter, it is that enforcement goes hand in hand with management.[69] Hence, the establishment of a well-functioning and balanced combination of management and enforcement mechanisms will likely be a key to success for the climate change non-compliance regime.

## Notes

1 This chapter has benefited greatly from rounds of comments from the project group and various external reviewers. Special thanks to the editors, Olav Schram Stokke, Jon Hovi and Geir Ulfstein. In addition, very helpful comments have been received from Georg Børsting and Sebastian Oberthür. Lynn P. Nygaard's comments and language corrections have improved the chapter considerably.

2 *ENDS Report* (2001) 'Kyoto Protocol Rules Finalised', November, no 322, 53.

3 Issues related to the reporting and verification of land-use emissions and sinks were not settled in Marrakesh.

4 See Werksman in this volume.

5 The relationship between institutional design and problem characteristics is further discussed in Wettestad (1999).

6 Thanks to Lynn Nygaard for pointing out this aspect.

7 The ECE has a membership of 55 states (see www.unece.org). Both the US and Canada are members of both the ECE and CLRTAP.

8 For instance, many parties to the 1985 Sulphur Protocol had already achieved the 30 per cent reduction target upon signing the Protocol. See, e.g., Underdal and Hanf (2000). For an assessment of the more recent 1999 Gothenburg Protocol, see Wettestad (2002).

9 Interview with Harald Dovland, Norwegian Ministry of Environment, 4 January 2002.

10 For more on the character of the ozone depletion problem and the related mitigation policies, see Wettestad in Miles et al (2001).

11 See, e.g., Parson and Greene (1995), and Parson (1996).

12 Exactly how benign depends upon the room for selling 'hot air' within the emissions trading regime.

13 This is Regulation 2037/2000, which replaced Regulation 3093/94. See Haigh (2002, p6.12). Thanks to Georg Børsting for pointing out this important difference.

14 See Oberthür (1997; 2001).

15 With regard to consumption, information is derived from reports on production, imports and exports, and substances destroyed. See Oberthür (1997; 2001).

16 For more specific information and assessments of the monitoring mechanism, see Hyvarinen (1999) and Haigh (2002).

17 The 1993 decision was modified in 1999, as described later in this chapter.

18 Harald Dovland, personal communication, 1991.

19 In this connection, see also di Primio (1996).

20 *Yearbook of International Co-operation on Environment and Development 2001/2002* (2001) Earthscan, London /The Fridtjof Nansen Institute, Lysaker, 94. The latest report was published in 2002 (WMO/UNEP, 2002).

21 Reuters (1999) 'UN Says Progress Made In Shielding Ozone', 30 November.

22 See, e.g., Tallberg (2002).

23 Interview with Lars Lindau, Swedish Environmental Protection Agency, 21 October 1999. Lindau has been Chairman of CLRTAP's Working Group on Abatement Techniques.

24 For instance, in 1996, only 5 per cent of the Parties reported on time. First Report of the Implementation Committee, EB.AIR/1998/4, 3.

25 First Report of the Implementation Committee, EB.AIR/1998/4, 3.

26 Interview with Harald Dovland, Norwegian Ministry of Environment, 4 January 2002. See also ECE Secretariat note (2001), 'Overview of Compliance Mechanisms Under ECE Environmental Conventions'.

27 EB.AIR/2001/3, (2001) 'The Fourth Report of the Implementation Committee', 3 October, 13.

28 *Green Globe Yearbook* (1996), Oxford University Press/The Fridtjof Nansen Institute, Lysaker, 107.

29 *Green Globe Yearbook* (1996), Oxford University Press/The Fridtjof Nansen Institute, Lysaker.

30 Interviews with Per Bakken, Norwegian Ministry of Environment, 18 October and 21 December 1995; Ivar Isaksen, University of Oslo, 12 December 1995.

31 'Report of the Twenty-Seventh Meeting of the Implementation Committee under the Montreal Protocol', 13 October 2001, UNEP/Ozl.Pro/ImpCom/27/4, 5.

32 Decision 93/389/EEC on the monitoring mechanism was formally amended on 5 April 1999, by Council Decision 1999/296/EC.

33 These seven countries were Luxembourg, Portugal, Italy, Ireland, Greece, Spain and Germany. See *Europe Environment* (2002) 'Air Quality: More Warnings for Non-respect of Legislation', 23 April, 610–5.

34 Directive 2003/87/EC of the European Parliament and of the Council of 13 October 2003 establishing a scheme for greenhouse gas emission allowance trading within the Community and amending Council Directive 96/61/EC, *Official Journal of the European Union*, Brussels, 25 October. See Christiansen and Wettestad (2003) for an assessment of this Directive. See also European Commission (2001) 'Proposal for a Directive of the European Parliament and of the Council Establishing a Framework for Greenhouse Gas Emissions Trading Within the European Community and Amending Council Directive 96/61/EC', European Commission, Brussels, COM(2001)581, 26.

[35] Directive 2003/87/EC of the European Parliament and of the Council of 13 October 2003 establishing a scheme for greenhouse gas emission allowance trading within the Community and amending Council Directive 96/61/EC, *Official Journal of the European Union*, Brussels, 25 October.

[36] ECE/CLRTAP (1999) 'Strategies and Policies for Air Pollution Abatement', UN/ECE, Geneva.

[37] As stated by Levy (1993, p91): 'Strategy and policy reviews are not intepreted; they are simply collated and published. There is no effort to ascertain whose measures place them in compliance with either specific protocols or broader norms ... Although no one is ever "cross-examined", states frequently make oral statements offering clarifications and emphasizing major points at EB meetings'.

[38] For more details on the functioning of the Implementation Committee, see Parson and Greene (1995); Greene (1996); Victor (1998).

[39] The procedure included general aims: to avoid complexity; to be non-confrontational (although an investigative process could be initiated either by other parties or the Secretariat); to be transparent; and to leave the decision making to the Meeting of the Parties. See Gehring (1994, pp315–9) and Szell (1995).

[40] *Environmental Policy and Law* (1989) 'Non-Compliance with Ozone Agreement', vol 19, 5.

[41] Regulatory committees favour Member States in the sense that such committees need to vote *for* a proposal made by the Commission, as opposed to advisory/management committees, where a negative vote is required to oppose a Commission proposal. See Hyvarinen (1999).

[42] See Art 3, para 3 in the VOC Protocol.

[43] 1994 Protocol, Art 7.

[44] See ECE/EB.AIR/53, annex III.

[45] Interview with Harald Dovland, 17 November 1995.

[46] Interview with Harald Dovland, 4 January 2002.

[47] First Report of the Implementation Committee, EB.AIR/1998/4, 2.

[48] First Report of the Implementation Committee, EB.AIR/1998/4, 8.

[49] ECE Secretariat (2001) 'Overview of Compliance Mechanisms Under ECE Environmental Conventions', 7.

[50] Report of the Tenth Meeting of the Parties to the Montreal Protocol, UNEP/OzL.Pro.10/9, 3 December 1998, 16.

[51] Report of the Thirteenth Meeting of the Parties to the Montreal Protocol, UNEP/Ozl.Pro.13/10, 26 October 2001, 32.

[52] However, although the parties to the Montreal Protocol have not changed the formal legal obligation of non-compliant CEIT countries, the Implementation Committee has established new schedules for coming into compliance. These schedules were presented in the form of 'benchmarks' by the CEITs concerned and negotiated/agreed with the Committee. They have basically replaced the formal reduction schedule in the deliberations of the committee (i.e., the latter now checks whether the benchmarks have been met rather than the formal phase-out schedule that has been missed anyway). The Global Environment Facility produced a study that helped develop the benchmarks in 1999. Communication with Sebastian Oberthür, November 2002.

[53] According to Oberthür (2000, p39): 'Since the early 1990s, the Global Environment Facility (GEF) has, in cooperation with the Implementation Committee ... provided assistance to CEIT countries to enable them to comply with the Protocol'.

[54] *Environmental Policy and Law* (1999) 'Measures to Strengthen the Work', vol 29, no 1, 9.

[55] *Environmental Policy and Law* (1999) 'Measures to Strengthen the Work', vol 29, no 1, 9.

56 Report of the Twenty Seventh Meeting of the Implementation Committee under the Montreal Protocol, UNEP/Ozl.Pro/ImpCom/27/4, 13 October 2001, 3.

57 Report of the Thirteenth Meeting of the Parties to the Montreal Protocol, UNEP/Ozl.Pro.13/10, 26 October 2001, 31.

58 *International Environment Reporter* (2001) 'Implementation Issues Top Agenda of Annual Montreal Protocol Ozone Meeting', 24 October, 908.

59 Communication with Georg Børsting, Norwegian Ministry of Environment, September 2002.

60 ECE Secretariat (2001), 'Overview of Compliance Mechanisms under ECE Environmental Conventions', 8.

61 EB.AIR/2001/3 (2001) 'The Fourth Report of the Implementation Committee', 3 October.

62 ECE Secretariat (2001), 'Overview of Compliance Mechanisms Under ECE Environmental Conventions'.

63 ECE Secretariat (2001), 'Overview of Compliance Mechanisms Under ECE Environmental Conventions', 3–4.

64 Greece was fined in 2000 for not implementing waste directives.

65 European Commission (2001) 'Proposal for a Direcive of the European Parliament and of the Council Establishing a Framework for Greenhouse Gas Emissions Trading within the European Community and Amending Council Directive 96/61/EC', European Commission, Brussels, COM(2001)581, 14.

66 However, Expert Review Teams are an integral part of the operation of institutions such as the OECD.

67 However, as noted in the previous section, this issue is quite complex and touchy one in practice.

68 As noted by Mitchell in Chapter 3 of this book, the stakes in relation to the more short-term Kyoto targets are not that high, because the regime's definitions of compliance and flexibility make compliance easy and relatively inexpensive.

69 This is also the central message in Tallberg (2002). To some extent, the message has already been acted upon by climate negotiators, as there is a certain element of compliance management in the form of the Compliance Action Plan to be produced for 'review and assessment' by the Enforcement Branch. Communication with Sebastian Oberthür, November 2002.

## References

Christiansen, A. and Wettestad, J. (2003) 'The EU as a Frontrunner on Greenhouse Gas Emissions Trading: How Did It Happen and Will the EU Succeed?', *Climate Policy*, vol 3, 3–18

Climate Network Europe (2002) 'Emissions Trading in the EU: Let's See Some Targets', CNE comments on Commission emissions trading paper, www.climnet.org/EUenergy/ET.html, 5 March

di Primio, J. (1996) 'Monitoring and Verification in the European Air Pollution Regime', Working Paper 96-47, IIASA, Laxenburg

Gehring, T. (1994) *Dynamic International Regimes: Institutions for International Environmental Governance*, Peter Lang Verlag, Berlin

Greene, O. (1996) 'The Montreal Protocol: Implementation and Development in 1995' in Poole, J. and Guthrie, R. (eds) *Verification 1996: Arms Control, Environment and Peacekeeping*, Westview Press, Boulder, CO, 407–26

Haigh, N. (1996) 'Climate Change Policies and Politics in the European Community' in O'Riordan, T. and Jäger, J. (eds) *Politics of Climate Change: A European Perspective*, Routledge, London and New York, 155–86

Haigh, N. (ed 2002) *Manual of Environmental Policy: The EC and Britain*, Cartermill Publishing/IEEP, London

Hyvarinen, J. (1999) 'The European Community's Monitoring Mechanism for $CO_2$ and other Greenhouse Gases: The Kyoto Protocol and other Recent Developments', *Reciel*, vol 8, no 2, 191–7

Levy, M. (1993) 'European Acid Rain: The Power of Tote Board Diplomacy' in Haas, P. M., Keohane, R. O. and Levy, M. A. (eds) *Institutions for the Earth*, MIT Press, Cambridge, MA and London, England, 75–133

Miles, E. L., Underdal, A., Andresen, S., Wettestad, J., Skjærseth, J. B. and Carlin, E. M. (2001) *Environmental Regime Effectiveness: Confronting Theory with Evidence*, MIT Press, Cambridge, MA

Oberthür, S. (1997) *Production and Consumption of Ozone-Depleting Substances 1986–1995*, Deutsche Gesellschaft fur Technische Zusammenarbeit, Berlin

Oberthür, S. (2000) 'Ozone Layer Protection at the Turn of the Century: The Eleventh Meeting of the Parties', *Environmental Policy and Law*, vol 30, no 1, 34–41

Oberthür, S. (2001) *Production and Consumption of Ozone Depleting Substances 1986–1999: The Data Reporting System under the Montreal Protocol*, Deutsche Gesellschaft fur Technische Zusammenarbeit, Berlin.

Parson, E. A. (1996) 'International Protection of the Ozone Layer' in Bergesen, H. O., Parmann, G. and Thommessen, Ø. B. (eds) *Green Globe Yearbook*, Oxford University Press/The Fridtjof Nansen Institute, Oxford/Lysaker, 19–28

Parson, E. A. and Greene, O. (1995) 'The Complex Chemistry of the International Ozone Agreements', *Environment*, vol 37, no 2, 16–20; 35–43

Sand, P. (1990) 'Regional Approaches to Transboundary Air Pollution' in Helm, J. (ed) *Energy: Production, Consumption and Consequences*, National Academy Press, Washington, 246–263.

Skjærseth, J. B. (1993) *The Climate Policy of the EC: Too Hot to Handle?* EED Report 1993/2, Fridtjof Nansen Institute, Lysaker

Szell, P. (1995) 'The Development of Multilateral Mechanisms for Monitoring Compliance' in Lang, W. (ed) *Sustainable Development and International Law*, Graham and Trotman, London, 97–109

Tallberg, J. (2002) 'Paths to Compliance: Enforcement, Management, and the European Union', *International Organization*, vol 56, no 3, 609–43

Underdal, A. and Hanf, K. (eds 2000) *International Environmental Agreements and Domestic Politics: The Case of Acid Rain*, Ashgate, Aldershot

WMO/UNEP (2002) *Scientific Assessment of Ozone Depletion: 2002*, Global Ozone Research and Monitoring Project, Report no. 47, WMO, Geneva.

Victor, D. G. (1998) 'The Operation and Effectiveness of the Montreal Protocol's Non-Compliance Procedure' in Victor, D. G., Raustiala, K. and Skolnikoff, E. B. (eds) *The Implementation and Effectiveness of International Environmental Commitments*, MIT Press, Cambridge, MA, 137–77

Wettestad, J. (1999) *Designing Effective Environmental Regimes: The Key Conditions*, Edward Elgar, Cheltenham

Wettestad, J. (2000) 'The Complicated Development of EU Climate Policy', in Grubb, M. and Gupta, J. (eds) *Climate Change and European Leadership*, Kluwer Academic Publishers, Dordrecht, 25–47

Wettestad, J. (2001) 'The Vienna Convention and Montreal Protocol on Ozone Layer Depletion' in Miles, E. L., Underdal, A., Andresen, S., Wettestad, J., Skjærseth, J. B. and Carlin, E. M. (eds) *Environmental Regime Effectiveness: Confronting Theory with Evidence*, MIT Press, Cambridge, MA, 197–223

Wettestad, J. (2002) *Clearing the Air: European Advances in Tackling Acid Rain and Atmospheric Pollution*, Ashgate, Aldershot

Wiser, G. M. (1999) *Compliance Systems Under Multilateral Agreements: A Survey for the Benefit of Kyoto Protocol Policy Makers*, Center for International Environmental Law, Washington, DC

# Epilogue: The Future of Kyoto's Compliance System

Jon Hovi

The first commitment period of the Kyoto Protocol expires in 2012. Many observers expect that a new agreement, designed for the second commitment period, will then have seen the light of day. To what extent are the conclusions reached in this book likely to be valid for this new agreement? Needless to say, it is hazardous to speculate on events that lie almost a decade ahead. Still, a few comments on the subject are appropriate.

Two main scenarios may be envisioned. The first is that the Kyoto Protocol enters into force and works fairly well, and that current non-parties (notably the US) join in before 2012. Clearly, this is a goal that the parties to the Kyoto Protocol have set for themselves. It should be emphasized, however, that including more parties will not *necessarily* make the agreement more effective. As pointed out by Barrett (1999; 2002; 2003), one may have to choose between a treaty that is 'broad but shallow' and one that is 'narrow but deep'. In other words, broadening the set of Member States may be possible only by making the agreement less ambitious (e.g., in terms of targets for emissions reductions). Nevertheless, should this first scenario materialize, it is reasonable to expect the agreement for the next commitment period to share many of the characteristics of the Kyoto Protocol. In particular, the basic Kyoto strategy, with a strong focus on quantitative emissions permits and flexibility mechanisms, may be expected to survive beyond 2012. Moreover, the compliance system of the new agreement may be expected to follow the same lines as the one provided by the Marrakesh Accords. If this scenario materializes, therefore, we would largely expect the argument of this book to be valid for the next climate agreement as well.

The second main scenario is that before the end of 2012, the parties reject the Kyoto strategy and turn to some alternative way of mitigating climate change. This might happen if major weaknesses of the Kyoto strategy are revealed during the first commitment period, or if the US continues to object to the Kyoto model, while signalling a more sympathetic attitude to some other basic design for an international climate regime.

At least four major alternatives to the Kyoto system have been discussed in the literature (Müller et al, 2001; Lisowski, 2002).[1] The first is to use *intensity targets* that focus on greenhouse gas (GHG emissions) per unit of GDP, so that emissions are allowed to increase with economic growth. The proponents of this option argue that, because of the strong link between emissions and growth, intensity is a better measure of performance than the absolute level of emissions. Intensity targets are also likely to be an attractive model for developing countries, as this approach makes it feasible to improve climate performance even in the midst of a rapid economic transition. It has been proposed to link intensity targets to so-called 'best practice' levels. This means that the intensity target of a given country A would equal the intensity level of the party with the best climate practice (i.e., the lowest intensity level) of all parties comparable to A in terms of development. This system has the attractive property that emissions per capita tend to converge to best practice levels at all stages of development. Any remaining differences would thus be due to economic structure (Lisowski, 2002).

A second alternative is to use a *technology-focused approach.* Proponents of this alternative typically argue that constraints on emissions should be deferred, or even abandoned altogether, in favour of an approach focusing on technological development and technological standards. An underlying motivation is that climate change is a long-term problem, and that costs may be cut considerably by postponing restrictions on emissions until better technology becomes available. New technology could be imposed through mandatory requirements, such as a requirement that new fossil power plants installed after, say, 2020 must remove all carbon from their exhaust stream and use new synthetic fuels to capture and dispose of carbon released in the conversion process (Edmonds and Wise, 1999). In this way, new technology would be introduced gradually as old plants become due for replacement. This is likely to cut costs considerably. Also, technological innovation might improve fuel efficiency and help develop alternatives to carbon-emitting technologies. As in the case of intensity targets, a technology-based approach might help to motivate developing countries to participate, especially if participation is linked to transfers of new technology.

A third option is *price caps* allowing governments to allocate emissions permits to domestic industry beyond the original allowances if emissions trading drives the price of permits above a pre-determined level (a 'trigger price'). In effect, producers have the choice of either obtaining a permit in the market or buying a permit from the government at the pre-determined trigger price (Pizer, 2002). In this way, the market price is prevented from climbing above the trigger, meaning that a maximum price per unit of emissions is established. This ensures that the maximum cost of meeting a commitment is known in advance, both at the country level and for producers at the subnational level. Price caps might thus satisfy Washington's concern that costs under the Kyoto system could turn out to be unacceptably high.

Finally, *coordinated taxes* represent a fourth approach: imposing an additional cost on emissions of GHGs to provide an incentive to cut emissions beyond the level that would otherwise have been reached. Taxes promote economic efficiency by ensuring that the resulting emissions level is achieved at the lowest possible cost.[2] This requires, however, that taxes are coordinated internationally, so that emitters face the same tax everywhere (Cooper, 1998). Like price caps, a tax also makes the costs of reducing emissions more predictable than they are likely to be under a regime based on fixed emissions levels.

Clearly, the challenges for compliance generated by these alternative approaches differ significantly from those raised by the Kyoto Protocol. There are at least two reasons for this. First, the overall level of non-compliance could well be lower if an agreement for the second commitment period is based on one of the above approaches. Compared to the Kyoto regime, several of the above alternatives share the advantage that the costs of reducing emissions are likely to be more predictable. Predictability curbs non-compliance by reducing the risk that signatories accept commitments that they are in fact unwilling – or even unable – to fulfil.

Second, the difficulties connected to verification and enforcement are likely to differ from those encountered by the Kyoto strategy. On the one hand, some alternative approaches might prove at least as difficult to enforce. For example, this could be the case for a regime based on intensity targets. As in the Kyoto Protocol, a basic element would be a set of emissions targets. However, allowing such targets to depend on economic output implies additional challenges. One such challenge is that intensity growth rates are sensitive to the choice of economic output measure, and to economic upturns and downturns. This could potentially trigger disputes over whether or not a given country's target has actually been reached, and whether sufficient good faith efforts have been made to achieve the target. Moreover, to assess compliance it may be necessary to verify the underlying economic indicators themselves.

Other options might turn out to be easier to enforce than the Kyoto strategy. For example, this seems like a plausible hypothesis for a technology-based approach, which might enhance compliance through network externalities and technological lock-in (Aldy et al, 2003). Such an approach would also make it relatively easy to verify that each party has fulfilled its obligations (Barrett, 2003). It is easier to establish the nature of new power plants, or the emissions of new cars, than it is to measure a given country's total emissions of GHGs.

The literature on alternatives to the Kyoto strategy has so far been more concerned with economic efficiency than with compliance. Hence, the brief comments above should be seen as indicative at best. While this book has focused on conditions for compliance with the Kyoto Protocol, the corresponding conditions for compliance with other models for a climate regime largely remain a challenge for future research.

## Notes

1 Others distinguish a larger number of alternatives. Aldy, Barrett and Stavins (2003) review the Kyoto Protocol and 13 alternative 'policy architectures' for addressing the threat of global climate change. Bodansky (2003) considers six alternative options to the Kyoto cap-and-trade model.

2 The merits of taxes in terms of efficiency are particularly great when compliance costs are uncertain (e.g., Newell, Pizer and Zhang, 2003).

## References

Aldy, J. E., Barrett, S. and Stavins, R. (2003) 'Thirteen Plus One: A Comparison of Global Climate Policy Architectures', *Climate Policy*, vol 3, no 4, 373–97

Barrett, S. (1999) 'A Theory of Full International Cooperation', *Journal of Theoretical Politics*, vol 11, no 4, 519–41

Barrett, S. (2002) 'Consensus Treaties', *Journal of Institutional and Theoretical Economics*, vol 158, no 4, 529–47

Barrett, S. (2003) *Environment and Statecraft,* Oxford University Press, Oxford

Bodansky, D. (2003) *Climate Commitments: Assessing the Options*, www.pewclimate.org/global-warming-in-depth/all_reports/beyond_kyoto/index.cfm

Cooper, R. (1998) 'Towards a Real Treaty on Global Warming', *Foreign Affairs*, vol 77, no 1, 66–79

Edmonds, W. and Wise, M. (1999) 'Exploring a Technology Strategy for Stabilising Athmospheric $CO_2$' in Carraro, C. (ed) *International Environmental Agreements on Climate Change*, Kluwer, Dordrecht

Lisowski, M. (2002) 'The Emperor's New Clothes: Redressing the Kyoto Protocol', *Climate Policy*, vol 2, no 2, 161–77

Müller, B., Michaelowa, A. and Vrolijk, C. (2001) *Rejecting Kyoto: A Study of Proposed Alternatives to the Kyoto Protocol*, www.climate-strategies.org/rejectingkyoto2.pdf

Newell, R., Pizer, W. and Zhang, J. (2003) 'Managing Permit Prices to Stabilize Markets', http://hkkk.fi/~econ/research/Richard.pdf

Pizer, W. A. (2002) 'Combining Price and Quantity Controls to Mitigate Global Climate Change', *Journal of Public Economics*, vol 85, no 3, 409–34

# Index